Improving Indicators of the Quality of Science and Mathematics Education in Grades K–12

Richard J. Murnane and Senta A. Raizen, Editors

Committee on Indicators of Precollege
Science and Mathematics Education

Commission on Behavioral and Social Sciences and Education

National Research Council

NATIONAL ACADEMY PRESS
Washington, D.C. 1988

National Academy Press • 2101 Constitution Avenue, N.W. • Washington, D. C. 20418

NOTICE: The project that is the subject of this report was approved by the Governing Board of the National Research Council, whose members are drawn from the councils of the National Academy of Sciences, the National Academy of Engineering, and the Institute of Medicine. The members of the committee responsible for the report were chosen for their special competences and with regard for appropriate balance.

This report has been reviewed by a group other than the authors according to procedures approved by a Report Review Committee consisting of members of the National Academy of Sciences, the National Academy of Engineering, and the Institute of Medicine.

The National Academy of Sciences is a private, nonprofit, self-perpetuating society of distinguished scholars engaged in scientific and engineering research, dedicated to the furtherance of science and technology and to their use for the general welfare. Upon the authority of the charter granted to it by the Congress in 1863, the Academy has a mandate that requires it to advise the federal government on scientific and technical matters. Dr. Frank Press is president of the National Academy of Sciences.

The National Academy of Engineering was established in 1964, under the charter of the National Academy of Sciences, as a parallel organization of outstanding engineers. It is autonomous in its administration and in the selection of its members, sharing with the National Academy of Sciences the responsibility for advising the federal government. The National Academy of Engineering also sponsors engineering programs aimed at meeting national needs, encourages education and research, and recognizes the superior achievements of engineers. Dr. Robert M. White is president of the National Academy of Engineering.

The Institute of Medicine was established in 1970 by the National Academy of Sciences to secure the services of eminent members of appropriate professions in the examination of policy matters pertaining to the health of the public. The Institute acts under the responsibility given to the National Academy of Sciences by its congressional charter to be an adviser to the federal government and, upon its own initiative, to identify issues of medical care, research, and education. Dr. Samuel O. Thier is president of the Institute of Medicine.

The National Research Council was organized by the National Academy of Sciences in 1916 to associate the broad community of science and technology with the Academy's purposes of furthering knowledge and advising the federal government. Functioning in accordance with general policies determined by the Academy, the Council has become the principal operating agency of both the National Academy of Sciences and the National Academy of Engineering in providing services to the government, the public, and the scientific and engineering communities. The Council is administered jointly by both Academies and the Institute of Medicine. Dr. Frank Press and Dr. Robert M. White are chairman and vice chairman, respectively, of the National Research Council.

Library of Congress Cataloging-in-Publication Data
National Research Council (U.S.). Committee on Indicators of Precollege Science and Mathematics Education. Commission on Behavioral and Social Sciences and Education.

Bibliography
Includes index.
   1. Science—Study and teaching—United States—Evaluation. 2. Mathematics—Study and teaching—United States—Evaluation. I. Murnane, Richard J. II. Raizen, Senta A. III. National Research Council (U.S.). Committee on Indicators of Precollege Science and Mathematics Education.
Q183.3.A1I48 1987   507'.1073—dc19   87-31230
ISBN 0-309-03740-9

Printed in the United States of America

COMMITTEE ON INDICATORS OF PRECOLLEGE SCIENCE AND MATHEMATICS EDUCATION

RICHARD J. MURNANE (*Chair*), Harvard University (economics)
LLOYD BOND, University of Pittsburgh (psychometrics)
NORMAN O. FREDERIKSEN, Educational Testing Service, Princeton, New Jersey (psychology)
ALICE B. FULTON, University of Iowa (biochemistry)
GERALD HOLTON, Harvard University (physics, history of science)
LYLE V. JONES, University of North Carolina (psychometrics)
C. THOMAS KERINS, Illinois State Board of Education, Springfield (education administration)
GEORGE MILLER, University of California, Irvine (chemistry)
HAROLD NISSELSON, Westat, Inc., Rockville, Maryland (mathematical statistics)
JEROME PINE, California Institute of Technology (physics)
MARY BUDD ROWE, University of Florida (science education)
MARSHALL S. SMITH, Stanford University (education, measurement, and evaluation)
WAYNE W. WELCH, University of Minnesota (science education)
SAMUEL J. MESSICK (ex officio), member, National Research Council Committee on Research in Mathematics, Science, and Technology Education; Educational Testing Service, Princeton, New Jersey (psychometrics, cognitive science)

SENTA A. RAIZEN, *Study Director*
ROLF BLANK, *Research Associate*

Preface

This is the second report of the Committee on Indicators of Precollege Science and Mathematics Education. The committee was established by the National Research Council to develop better indicators of the condition of science and mathematics education in the nation's schools. The impetus for the work came from a convocation held by the National Academy of Sciences in spring 1982 on mathematics and science education; additional motivation came from various reports on the condition of education that appeared in fall 1982 and spring 1983, including those by the National Science Foundation (NSF), the National Commission on Excellence in Education, and the Twentieth Century Fund.

These reports found serious inadequacies in precollege education; a number of them suggested that many U.S. students leave high school without adequate preparation in science and mathematics, whether for the job market, for continuing their education, or for informed citizenship. The reports identified such specific school deficiencies as teacher shortages, inadequate curricula, and low standards of student performance. These reports elicited widespread concern about the state of schooling; however, questions were also raised about the quality of the information used to formulate many of the conclusions and policy recommendations in the reports.

This concern led to the creation of our committee, which is charged with laying a foundation for the development of an adequate monitoring system for use at the national, state, and local levels, so that the condition of mathematics and science education can be tracked, particularly the effects of current efforts at improvement.

The committee's first report, issued in 1985, concentrated on conventional indicators and current data bases. Both the committee and reviewers suggested that the next step should be consideration of new indicators that would provide more penetrating insights on the condition of science and mathematics education. Hence, with support from the National Science Foundation, the committee has continued its work on developing an improved system of indicators, including recommendations on more imaginative assessment measures for present use as well as on research to create new indicators. Some of the problems identified in the first report have received further attention, for example, defining teaching effectiveness, developing indicators of the quality of curriculum content, and improving assessment of student performance. Some potential indicators identified but not selected for discussion in the first report have been reexamined.

The committee conducted several activities to enlarge the perspectives and knowledge on indicators represented by its members. In fall 1985, some 50 outside experts participated in a workshop chaired by Lyle Jones aimed at developing improved approaches to indicators (Appendix A is a list of participants). Discussions in each of the subgroups led to the organization of the report into sections dealing with scientific and mathematical literacy; assessment of student learning; measures of student behaviors, attitudes, and motivation; measures of teaching effectiveness; assessment of the quality of curriculum content; and indicators of financial investment. On the last topic, two papers were commissioned, one by Ward S. Mason on indicators of federal investment in precollege science, mathematics, and technology education, and one by Kern Alexander on costs of school mathematics and science programs. Ideas suggested at the workshop on each of the six areas were subsequently used by the committee in formulating its report.

The committee also conducted a review of the science content in nine selected achievement tests used by many secondary schools. The results of the review, performed by a group of scientists and science teachers, were used by the committee in developing its findings on problems with achievement tests and in recommending strategies for improving methods of assessing learning in science. Appendix B summarizes the procedures and results of the test assessment and lists participants.

In spring 1986, the committee convened a meeting with representatives from state education agencies and another meeting with representatives of large local school districts. Each of the meetings was attended by administrators responsible for research and evaluation and by curriculum supervisors. The purpose of the meetings was to learn of the needs, interests, and concerns of education officials at these levels about indicators of science and mathematics education and to have them comment on the feasibility and usefulness of the committee's proposed approaches. The discussions with state and local officials, summarized in Appendix C, produced insightful comments on the committee's initial ideas as well as useful new suggestions of strategies for improving indicators. A number of these suggestions are reflected in this report.

In addition to its own activities, the committee's work has profited from several other efforts proceeding concurrently to develop and improve indicators of the quality of American education. The committee has kept in close contact with these efforts, which are described in Appendix D, and has both learned from them and influenced their work.

In formulating this report, the committee brought to this wealth of information the distinct disciplinary perspectives of its members. While the general findings and recommendations belong to the committee as a whole, I am appreciative of the hard work done by individual members in drafting the text for each chapter: by Alice Fulton and myself for Chapter 2, by Lloyd Bond and Harold Nisselson for Chapter 3, by Norman Frederiksen and Jerome Pine for Chapter 4, by Mary Budd Rowe and Wayne Welch for Chapter 5, by George Miller and myself for Chapter 6, by Marshall Smith and Senta Raizen for Chapter 7, and by C. Thomas Kerins for the sections in several chapters dealing with implications for state education agencies. The staff drafted Chapter 8 and Appendixes B, C, and D, and Harold Nisselson drafted Appendix E.

The committee is grateful for the support provided by the National Science Foundation for its work and the unfailing help and encouragement extended by NSF's Richard Berry. We would also like to acknowledge the contributions to the committee's early deliberations by John Truxal (State University of New York, Stony Brook), chair of the committee in 1985, Thomas Lippincott (University of Wisconsin), and Henry Pollak (Bell Communications Research, Inc.), all three of whom were committee members who had to resign midway through our work. Special appreciation is due to F. Joe Crosswhite (National Council of Teachers of Mathematics), John Dossey (Illinois State University), Henry Pollak, and Thomas Romberg (University of Wisconsin), all of whom reviewed and commented on the text of the report with respect to mathematics.

I would particularly like to thank Senta Raizen, the study director of this project, for her enormous contribution. Not only did her knowledge, wisdom, wit, and hard work contribute directly to the report, but she also motivated all committee members to work much harder on this project than they initially had in mind. The committee also appreciates the work of staff members Rolf Blank and Barbara Darr and the careful editing of Christine McShane.

RICHARD J. MURNANE, *Chair*
Committee on Indicators of Precollege
Science and Mathematics Education

Contents

1	Summary and Recommendations	1
2	Indicators of Science and Mathematics Education	15
3	What Are Indicators?	27
4	Indicators of Learning in Science and Mathematics	40
5	Indicators of Student Behavior	73
6	Indicators of Teaching Quality	90
7	Indicators of Curriculum	119
8	Indicators of Financial and Leadership Support	143
	References	152

APPENDIXES

A	Colloquium on Indicators of Precollege Science and Mathematics Education: Participants	171
B	Review of Science Content in Selected Student Achievement Tests	175
C	Summaries of Meetings with Representatives of State and Local Education Agencies	181
D	Current Projects on Indicators	197
E	Coordination of Strategies for Collecting Data	206
	Index	209

1

Summary and Recommendations

The Committee on Indicators of Precollege Science and Mathematics Education was established by the National Research Council to develop indicators of the condition of science and mathematics education in the nation's schools. The committee's first report concentrated on conventional indicators. In this report, the committee makes recommendations for improved ways of monitoring the condition of education in these critical fields. Our recommendations are based on two premises:

- All students need the knowledge and reasoning skills that good science and mathematics education provides. Not only should students leave school scientifically and mathematically literate, but they should also have acquired the mental tools with which they can renew that literacy throughout their lives.
- What teachers and students do in schools determines how much learning takes place. Student and teacher behaviors are influenced by a variety of incentives and constraints. Among the many influences on behavior are curriculum mandates and curricular support materials, working conditions for teachers, and resources at the classroom level. This simplified model of the educational process provides the framework for our report.

This chapter presents the committee's recommendations. Support for the recommendations, definitions of what the committee means by *high-quality education* and by the term *indicators*, and caveats about the interpretation of indicators are provided in the subsequent chapters.

INDICATORS OF SCIENCE AND MATHEMATICS EDUCATION

The committee makes three kinds of recommendations. We recommend a number of key indicators, to which we assign the highest priority. We recommend supplementary indicators, which, although of lesser importance, would also improve knowledge of the quality of mathematics and science education. The key indicators we recommend are listed below, and the supplementary indicators are listed on page 4. We also make a number of proposals for research, either to validate the recommended indicators or to lead to the development of additional indicators.

For two of the key types of indicators—dealing with assessment of student learning and assessment of curriculum content—important development work needs to be done before they can become useful indicators. For the assessment of learning in mathematics and science, tests and exercises need to be developed that will allow assessment of conceptual knowledge, process skills, and higher-order thinking in addition to the factual knowledge and skills assessed by tests in current use. For the assessment of curriculum, exemplary frameworks containing substantive content and desirable learning goals need to be constructed, each spanning several grade levels, to provide referents against which textbooks and other curriculum components can be evaluated.

Recommended Key Indicators

- Extent of student learning in mathematics and science
- Extent of scientific and mathematical literacy of adults
- Enrollment data for mathematics and science courses taken by students in high school and the amount of time spent on the study of science and mathematics in elementary and middle/junior high school
- Nature of student activities during science and mathematics instruction

- Extent of teachers' knowledge in the subject matter that they are expected to teach
- Salaries paid to college graduates with particular subject-matter specialties who choose to enter various occupations
- Quality of the curriculum content in state guidelines, textbooks and associated materials, tests, and actual classroom instruction in science and mathematics through matching to exemplary curriculum frameworks along four dimensions: breadth and depth of treatment and scientific and pedagogic soundness

For each key indicator, unless otherwise noted in a specific recommendation, data should be collected in four-year cycles. For indicators dealing with student learning, student behavior, teaching effectiveness, and quality of the curriculum, information should be collected and analyzed so it can be reported by student subgroup—that is, data should be aggregated not only by region (or state, if current efforts in that direction proceed), but also by students' age or grade level, gender, race, ethnicity, socioeconomic status, and type of community (urban, suburban, rural). The reason for aggregating by student demographic variables is to establish to what extent there are systematic inequities in the distribution of resources devoted to science and mathematics education and systematic differences in student learning.

Some of the proposed indicators are most appropriate at the national level, for example, assessment of the scientific and mathematical literacy of the general population. Others may be most policy-relevant at the school level, for example, the information obtained by observing student activities during instruction. Still others are relevant at the national, state, and local levels, for example, assessments of student learning or teacher knowledge of subject matter. When appropriate, the recommendations note the policy level for which an indicator is intended. Recommendations for research are addressed both to the research community and to those funding agencies concerned with better understanding and monitoring of science and mathematics education.

> **Recommended Supplementary Indicators**
> - Amount of time spent on science and mathematics homework
> - Teacher preparation—college courses in mathematics and science, majors and minors, advanced degrees
> - Teachers' use of time outside the classroom spent on professional activities related to their teaching of mathematics and science
> - Materials, facilities, and supplies available and used by teachers in mathematics and science instruction
> - Level of federal financial support for science and mathematics education
> - Commitment of resources by scientific bodies for the improvement of mathematics and science education in the schools

In the chapters that follow, our recommendations for improving indicators in current use and developing new ones appear within the appropriate domain of science and mathematics education: student learning (Chapter 4) and student behavior (Chapter 5), teaching quality (Chapter 6), quality of the curriculum (Chapter 7), and financial and leadership support (Chapter 8). A consequence of this organization is that recommendations for key indicators are intermingled with recommendations for supplementary indicators and with recommendations for research.

The remainder of this chapter presents all the recommendations with amplifying material as they appear in the report, spelling out the recommended key and supplementary indicators in greater detail as well as needed research. The recommendations are organized by domain of mathematics and science education, as they are in the chapters of this report.

RECOMMENDATIONS

Indicators of Learning in Science and Mathematics

Indicators of student learning at the national, state, and local levels should be based on scores on tests that are consonant with the curriculum and all major curricular objectives, including the learning of factual and conceptual knowledge, process skills, and higher-order thinking in specific content areas. Measuring progress toward this last objective is especially important, since it is possible that pressures on school practitioners to increase student scores on multiple-choice tests emphasizing recall of factual information may result in diminished attention paid to the development of higher-order thinking skills. In order to establish how well major curricular objectives are being met, test items used to assess students' mathematics and science learning should not be exclusively in a multiple-choice format. A significant number of items using an open-ended pencil-and-paper format and a hands-on problem-solving format should also be used.

Research and Development: To provide the requisite tests for use as indicators of student learning, the committee recommends that a greatly accelerated program of research and development be undertaken aimed at the construction of free-response materials and techniques that measure skills not measured with multiple-choice tests. The committee urges that the development of science tests at the K–5 level receive immediate attention.

Techniques to be developed include problem-solving tasks, as exemplified by the College Board Advanced Placement Tests; pencil-and-paper tests of hypothesis formulation, experimental design, and other tasks requiring productive-thinking skills, as exemplified by questions in the British Assessment of Performance Unit Series; hands-on experimental exercises, as exemplified by some test materials administered by the National Assessment of Educational Progress (NAEP) and the International Association for the Evaluation of Educational Achievement (IEA); and simulations of scientific phenomena with classroom microcomputers that give students opportunities for experimental manipulations and prediction of results.

The creation of new science tests for grades K–5 should be done by teams that include personnel from the school districts that have been developing hands-on curricula to ensure that the new tests match the objectives of this type of instruction. In addition to providing valid national indicators of learning in areas of great importance, such new assessment materials for science in grades K–5 will provide models of tests that state and local school officials may want to adopt and use.

Key Indicator: The committee recommends that assessment of student learning using the best available tests and testing methods continue to be pursued in order to provide periodic indicators of the quality of science and mathematics education.

Tests should be given to students in upper-elementary, middle, and senior high school (for example, in grades 4, 8, and 12). Because of the rapid changes taking place in science instruction in grades K-5, assessment at this level should be carried out every two years, using exercises developed according to the preceding recommendation. For higher levels, a four-year cycle is appropriate. The tests should be given to a national sample, using matrix-sampling techniques. Test scores should be available for each test item or exercise and should be reported over time and by student subgroups (e.g., gender, race, ethnicity, type of community). Similar procedures are appropriate for indicators of state or district assessments of student learning.

Research and Development: The committee recommends that a research and development center be established to provide for the efficient production, evaluation, and distribution of assessment materials for use as indicators of student learning at district, state, and national levels and for use by teachers in instruction.

The center should function as a centralized resource and clearinghouse that would make it possible for school people to survey the available assessment materials and obtain those desired. The center might be called the National Science and Mathematics Assessment Resource Center.

Key Indicator: The committee recommends that, starting in 1989, the scientific and mathematical literacy of a random sample of adults (including 17-year-olds) be assessed. The assessment should tap the dimensions of literacy discussed in Chapter 2 and should be carried out every four years.

To make the desired types of assessment possible, effort should be devoted over the next two years to developing interim assessment tools that use some free-response and some problem-solving components; these assessment tools should be used until more innovative assessment techniques are available. The data collected should be aggregated and reported by age, gender, race, ethnicity, socioeconomic status, and geographic region so as to establish to what extent there are systematic inequities in the distribution of scientific and mathematical literacy.

Indicators of Student Behavior

Key Indicator: The committee recommends that data on secondary school course enrollment be gathered on a four-year cycle for both mathematics and science. The specific data to be gathered are the number of semesters of science and mathematics taken by students and total enrollment in the variety of science and mathematics courses offered in secondary schools.

Courses should be identified as to level of difficulty (e.g., for eighth-grade mathematics: remedial, typical, enriched, algebra). The indicators to be constructed from these data are the average number of mathematics and science courses taken and the percentage of students enrolled in specific courses.

Key Indicator: The committee recommends that the data to be gathered at the elementary- and middle-school level, equivalent to course enrollment data, be the number of minutes per week devoted to the study of science and mathematics. The indicator should also be expressed both as a ratio of all instructional time and of total time spent in school.

At each policy level—national, state, and local—experts may wish to define the minimum amount of class time necessary in each grade, particularly for science. Because of the importance of possible differences among various groups (ethnic and racial, gender, socioeconomic status, etc.) we recommend that the data be collected at the level of both the school and the individual student.

Key Indicator: The committee recommends development of a time-use study involving external observers to obtain some indication of the quality of the science and mathematics instruction being received. In science classes, this would include, in addition to the teaching of conceptual and factual knowledge, the percentage of time spent by students involved in the processes of science (observing, measuring, conducting experiments, asking questions, etc.). A similar study is recommended for mathematics classes; a panel of mathematics educators should determine the nature of the student behaviors sought.

Supplementary Indicator: The committee recommends the collection of information on minutes per week spent on science and mathematics homework.

The frequency and detail necessary for gathering data on homework are the same as for in-school activities—that is, the information should be gathered every four years and allow analysis by ethnicity, race, gender, grade level, and size and type of community. National data are important for comparisons over time and with other countries; states and local districts may also wish to have this information. Care must be taken that homework done in school is not double counted as both homework time and instructional time.

Research and Development: The committee recommends further research and development on possible supplementary indicators in the following three areas of out-of-school behaviors, with the goal of clarifying their relationships to student mathematics and science learning:
- Amount of time (minutes) devoted to out-of-school science and mathematics activities, for example, going to zoos and science museums, watching science programs on television, reading science books, playing with a

computer at home, voluntarily doing science projects or mathematics puzzles.
- Percentage of students reporting that they use (apply) the concepts of science and mathematics from time to time in their own lives. One way to implement this indicator is to conduct a survey on the number of times students faced a personal decision and relied on something that they learned in science or mathematics to help them make that decision.
- Percentage of students reporting that they use the concepts of science and mathematics to help them address some persistent societal problem.

Research and Development: The committee recommends continued research on linkages between student learning and various student activities, on more effective ways of assessing activities that affect learning, and on the factors that influence individuals to engage in these activities.

Research and Development: Given the importance attached by science and mathematics educators to the development of attitudes that will foster continuing engagement with science and mathematics, the committee recommends that research be conducted to establish which attitudes affect future student and adult behavior in this regard and to develop unambiguous measures for those that matter most.

Research and Development: The committee recommends research to identify and validate constructs related to the continuing involvement of students and adults with science and mathematics throughout their lives. In addition to the refinement of these constructs, strategies should be explored for obtaining indicators of the relevant constructs and associated behaviors.

Indicators of Teaching Quality

Key Indicator: The committee recommends that samples of teachers be selected to take tests that probe the same content and skills that their students are expected to master. For this purpose, tests for teachers should be developed to include the same kinds of improvements that the committee recommends for tests of student learning.

The distribution of teachers' test scores should be reported by student background and characteristics. This will provide information about the distribution across different student groups of teachers who are in command of the mathematics and science they are expected to teach. Both current distribution and change over time are of interest; therefore, tests should be given every four years to a sample of all teachers and every two years to a sample of newly hired secondary school mathematics and science teachers.

Supplementary Indicator: The committee recommends reorganization of the information currently being collected on teacher preparation (college courses in mathematics and science, majors and minors, advanced degrees), using various student groups taught as the reporting groups of interest.

The information reported should display the percentage of students with particular backgrounds and characteristics who are being taught mathematics and science in elementary school as well as courses in these domains in secondary school by teachers with specific college preparation. For this indicator also, four-year cycles are appropriate for collection and analysis of information.

Research and Development: The committee recommends that research be undertaken on two issues: the impact of teachers' knowledge of subject matter on their effectiveness in teaching these subjects to students, and the role of early home and school experiences in determining the decisions to become a teacher and on how and what to teach.

Supplementary Indicator: The committee recommends that time-budget studies be conducted, asking teachers to record how they spend time related professionally to their present or future classroom activities, other than in the classroom itself, during a particular period, perhaps a week.

The information collected should be evaluated against sets of activities identified by experts as advancing effectiveness in the classroom in teaching mathematics or science. Investigations of the relationships between professional activities reported by teachers and teaching effectiveness should be conducted to help refine this indicator.

Research and Development: The committee recommends research on the following aspects of the behavior of teachers in science and mathematics instruction:

- the factors affecting teacher responses to changes in the intended curriculum;
- the use of hands-on experiences involving concrete materials, laboratory experiments, and computers; and
- allowing an adequate period of time for students to formulate responses to questions.

Supplementary Indicator: The committee recommends that data be collected on a four-year cycle through open-ended surveys on the materials, facilities, and supplies available and used by teachers in mathematics and science instruction.

An indicator can be constructed from this information by reporting on the levels of resources being used in the classroom by student subgroups of different backgrounds and competencies.

Key Indicator: The committee recommends collection at least every three years (preferably every two years) of detailed information on the salaries paid to college graduates with particular subject matter specialties who choose to enter various occupations.

The information should include data on starting salaries and on salaries after 15 years of experience. These data should be reported in a manner that facilitates comparisons of salaries in teaching with salaries in other occupations for college graduates trained in particular sciences and mathematics.

Indicators of Curriculum Quality

Research and Development: In order to develop indicators of breadth of content coverage in the science and mathematics curriculum, the committee recommends that exemplary frameworks be constructed for the following curriculum blocks: grades K–5 science, grades K–5 mathematics, grades 6–8 science, grades 6–8 mathematics, grades 9–12 literacy in science, grades 9–12 literacy in mathematics, grades 9–12 science for college-bound students, and grades 9–12 mathematics for college-bound students. The frameworks for grades K–5 and 6–8 science should be accorded the highest priority.

The frameworks must represent the structures of the subject matter and desirable learning goals, or alternatives among desirable goals.

Key Indicator: Once the frameworks are constructed, the committee recommends that three elements of the intended curriculum should be matched and rated against them for content coverage: state guidelines, textbooks and such associated materials as computer software and laboratory exercises, and tests. The frameworks should also be used to analyze the content coverage of the implemented curriculum (i.e., the content presented to the student as reported by classroom teachers).

The ratings obtained through analysis of the three elements of the intended curriculum and analysis of the implemented curriculum will provide the raw material for the construction of indicators of content coverage. The ratings should be carried out every four years at the national level in synchronization with the student assessments recommended above so that the indicators can be used together.

Research and Development: The committee recommends that research be carried out to establish the validity of teacher-reported information regarding content coverage in the classroom.

Research and Development: Standards of excellence should be developed based on the best of curricula in current use.

High-quality programs encompassing the curriculum blocks suggested above should be selected, profiled, and analyzed to provide models of excellence in depth of content coverage, scientific accuracy, and pedagogic soundness of science and mathematics curricula.

Key Indicator: The quality of the curriculum should be assessed by expert panels along three dimensions: depth of content treatment, scientific accuracy, and pedagogic soundness. Ratings for each of these quality dimensions should be assigned to the three elements of the intended curriculum (i.e., state guidelines, texts and associated materials, and tests). Assessments regarding depth of treatment should also be made of the implemented curriculum through teacher and student surveys and classroom observation.

To assess the depth of content treatment, the frameworks developed according to the recommendation made above should be used to identify the critical topics that constitute a coherent curriculum. Weights assigned by each rating panel regarding the depth of treatment desired for a given topic must be made explicit in reporting results.

The assessment of the scientific accuracy of the intended curriculum should be carried out by scientists in the relevant disciplines. The scientific content of the frameworks shoud be used to construct the tests of teacher competency of subject matter recommended in

Chapter 6 and such tests used as a minimum measure of the scientific accuracy of the actual curriculum experienced by students.

Research and Development: The committee recommends research to provide validity checks on the standards being used to assess depth of treatment, scientific accuracy, and pedagogic soundness of science and mathematics curricula.

For example, research should be undertaken to establish what pedagogic knowledge teachers need to have and need to know how to use in order to teach science and mathematics effectively to students of different ages, backgrounds, and competencies.

Indicators of Financial and Leadership Support

Supplementary Indicator: The committee recommends the construction of a set of accounts detailing the level and type of support for science and mathematics education from all departments and agencies of the federal government that fund relevant programs.

The importance of having reliable annual data on the level of federal financial support merits the investment necessary to construct such a set of accounts. Agencies should be encouraged to report budget and funding data by categories identifiable as precollege mathematics and science education, and funds should be made available (possibly through NSF) to perform the necessary analyses. The kind of disaggregation of financial support for science and mathematics education found in the NSF budget could be used as a model for developing the recommended cross-agency indicator of federal support.

Supplementary Indicator: The committee recommends that indicators be designed using budgetary data of scientific bodies and information on staff time and volunteer time devoted to education and that these indicators be routinely available to reflect the commitment of resources by scientific bodies for the improvement of mathematics and science education in the schools.

2
Science and Mathematics Education

American public schools are populated by 40 million students and 2 million teachers. The magnitude and diversity of American schooling make it impossible to deliver a complete and exact picture of how well the schools are doing: simplification and abstraction are necessary and inevitable. In attempting to abstract from the complexities of American schooling and to propose new indicators of the quality of science and mathematics education, the committee has taken on a task that requires a clear sense of definitions, methods, and goals. This chapter first defines what we take to be the purposes of science and mathematics education and then presents a conception of schooling that has shaped our choice of indicators.

SCIENTIFIC AND MATHEMATICAL LITERACY

The recognition that societies are changing rapidly is widespread—witness such terms as *information age, postindustrial society,* and *global economy* that have come into common usage. These changes are spurred by and give rise to the development of ever more powerful technologies. Several writers who have pondered the implications of these changes have made the point that the accelerating change that characterizes U.S. and other Western societies may well require a higher degree of scientific and mathematical literacy than ever before (Toffler, 1980; Naisbitt, 1982; Zarinnia and Romberg, 1986).

In general, the committee agrees with this view: we believe that all students should have the tools of knowledge and judgment that science and mathematics provide. These tools will enable them to prepare for careers and cope with rapid changes in tomorrow's labor markets, make informed choices in their private and family lives, and understand issues of broader scope. And all students, as they become adults, should be free to enjoy the enlargement of the world that comes with scientific and mathematical literacy. Clearly, then, not only should students leave school literate in science and mathematics, but they should also have acquired the mental tools with which they can renew that literacy throughout their lives.

Any useful attempt to determine to what extent schools are meeting this broad goal must reflect an informed view of what constitutes literacy in science and mathematics. The committee suggests that there are several dimensions of scientific and mathematical literacy, each of which needs to be addressed seriously. The following descriptions of these dimensions are intended to be normative: they are the ideals against which science and mathematics curricula and instruction should be compared. Because the sciences and mathematics, despite their many interconnections, are quite differently constituted, literacy in each of these domains is discussed separately.

*Literacy in Science**

The dimensions of scientific literacy that should be integral to any educational program include the nature of the scientific world view, the nature of the scientific enterprise, scientific habits of mind, and the role of science in human affairs. The first three of these dimensions deal with knowledge of science and intellectual skills; the fourth deals with the relation between science and society. Each is characterized in somewhat greater detail below.

The Nature of the Scientific World View Over the last three centuries or so, scientific activity in many fields has resulted in a set of interconnected and testable notions about the nature of the world

---

*We thank F. James Rutherford, American Association for the Advancement of Science, for suggesting the four dimensions of scientific literacy discussed in this section.

and its parts. While the details continue to change with time, these notions amount to a rather robust and useful construct of ideas that deserves the name *scientific world view*. The ideas that constitute its elements have various levels of complexity or abstraction.

- At the highest level of abstraction, there are grand conceptual schemes that bring together and bring order to large numbers of observations, concepts, and theories, each of which is of lesser generality and applies only to some narrower field of science. Some of these grand schemes can be expressed in words or numbers; others can only be understood mathematically. Examples include the Newtonian universe, organic evolution, and plate tectonics.
- At a more operational level, the scientific world view finds expression in theories and mathematical models that organize facts and laws in ways that help one understand a particular aspect of the world. Examples include gravitation, solid-state science, statistics, weather systems, and economic determinism.
- Particular concepts, mathematical tools, or techniques can reappear in various scientific specialties, thereby not only helping to suggest new advances but also providing syntheses among different parts of the total scientific world view. Examples of such recurring concepts or tools include scale, cycles, waves, estimation, energy, antibodies, and probability.

The scientific world view also consists of general beliefs that have shown their worth over time. These include the following notions:

- The world of phenomena is rationally understandable and not capricious; causal relations may be found.
- Good scientific theories permit deductions that can be checked against experience. This testing frequently takes the form of comparing measurements of the real world to numerical predictions.
- Individual data and observations are subject to some uncertainty, but phenomena are consistent.
- Throughout the history of scientific growth, and despite major advances along the way, scientists have often found a few fundamental thematic notions useful and motivating: for example, the search for unity or unification among diverse phenomena; the use of mechanical or mathematical models; and such notions as simplicity, parsimony, symmetry, evolution, causality, order or hierarchy, and continuity or discontinuity.

The Nature of the Scientific Enterprise The scientific enterprise subscribes to a set of value commitments in principle—an ethos of science—that can be explicitly formulated. For individuals to understand the conclusions that scientific research yields concerning the natural world, they must understand these principles and some of the characteristics of the way science works:

• Science is both theoretical and empirical, and these two aspects reinforce each other. Mathematical models without data are sterile; observations without measurement of some kind remain largely impressionistic.

• Science is not only a personal, individual calling but also a social activity carried out by individuals who collaborate over time and space. Some of this accessibility is a consequence of the universality of mathematics. This characteristic makes scientific activity, next to mathematics, one of the most international and shareable experiences and opens scientific research and teaching to all talents everywhere.

• The substance of science at any given time is found in the consensus among scientists, as reflected largely in current writings, data bases, and mathematical formulations. As scientific knowledge becomes more developed and inclusive, findings reached at any particular time are seen as tentative. As mathematical models attain greater generality, they provide novel tests of the current formulations. Thus, science is an enterprise concerned with discovery of the new and with testing (and correction when needed) of the old.

• Necessary conditions for understanding the processes of science include familiarity with a wide range of natural phenomena; asking questions and forming hypotheses; understanding the need for tests or controlled comparisons; embracing theories of measurement, evidence, and data; and accepting a method of notation or formalism, most often mathematical, that allows an unambiguous and replicable depiction of a set of phenomena.

• Pure scientific research often points the way to practical applications through engineering development, and the latter in turn can help make basic experimental science much more effective.

Scientific Habits of Mind Individuals, scientists or not, need to be capable of analytical thinking in the context of the various sciences and mathematics. This capability can be taken to include:

- Identifying questions or formulating hypotheses relative to a problem, recognizing when such questions ought to be quantitative, and being able to express them mathematically if that is appropriate.
- Identifying and seeking out information (numerical or otherwise) relevant to a problem or issue.
- Using that information to test the hypothesis or answer the question, while appreciating the limits placed by samples or intrinsic uncertainty.
- Playing with information, in the sense of solving puzzles and raising hypotheses.
- Offering arguments and counterarguments that can be tested by reference to data or accepted principles.
- Communicating, collaborating, and building consensus in order to develop a common language and common models.

The historic study of most scientific advances shows, however, that other, more individual, and even aesthetic, elements enter to assist the purely logical-critical faculty during the creative phase of scientific work. There is, in short, no single "scientific method." Moreover, accepting the scientific world view does not disqualify an individual from sensitivity to, or the appreciation of, artistic and humanistic achievements. In fact, one of the values of science is cultural. Participation in science can be satisfying in much the same way as participation in music and art.

Science and Human Affairs Science and mathematics are important to society because of their deep connections with and effects on human events and ideas. For example, mathematical concepts such as estimation, statistics, sampling, and risk underlie most public-policy concerns. The application of science and mathematics to such matters as health, industrial processes, agriculture, and the environment engage a large number of policy makers and individual citizens. From this it follows that, for a society to be scientifically literate, people need to understand some of the relationships between science, policy, and society.

- Tensions exist between science and society, because science must sometimes assert the presence of uncertainty and ambiguity when facts are needed by organizations that must make policy decisions or individuals who must make personal ones.
- Scientists may behave differently when they are involved in public policy decisions than when they are acting as researchers. The

ethics of science in public policy calls for objectivity, but this cannot always be expected in cases in which the self-interest of science itself is at stake.

• Science and technology have helped to better the human condition. But progress in science and technology can also have unanticipated negative effects on society. These effects can be monitored and in some instances modified by the action of alert citizens, provided they have the necessary educational background to obtain and interpret information on the ways in which science and technology influence personal, local, and national affairs.

• A historical sense of the way science grows can be useful. One cannot always predict what science will be of practical value in the future, because several seemingly unrelated lines of basic research may come together over time (sometimes decades) and contribute to breakthroughs. For example, cell culture was initiated to permit studies of development and other cellular behaviors; it subsequently made possible such medical advances as the polio vaccine. Much mathematics originally developed for its own beauty or interest has later provided tools for scientific or technological applications that were inconceivable earlier.

*Literacy in Mathematics**

The types of mathematical literacy—practical arithmetic, civic application, professional use of mathematics, and cultural appreciation—correspond roughly to the central objectives of the four hierarchical tiers in the educational system: primary, secondary, undergraduate, and graduate. Although it is useful to postulate levels of mathematical literacy corresponding to levels of education, some key elements are integral to all levels:

• Understanding the fundamental ideas of mathematics. For example, the Pythagorean theorem has theoretical and practical importance in all levels of mathematics learning, as does the notion of symmetry. These (and other) intrinsic mathematical ideas could provide benchmarks of literacy that transcend educational levels.

• Understanding the role of mathematics as the language of science and its role in describing the nature of complex systems.

---

*We thank Lynn Arthur Steen, St. Olaf College, for formulating the four levels of mathematical literacy discussed in this section.

Understanding that order can beget disorder (as in turbulence) and vice versa (as in statistical experiments); that mathematical models for growth can represent phenomena in biology, economics, and chemistry; and that mathematics is still being created to meet new needs are examples of perceptions about the nature of mathematics that should be part of mathematical literacy at every level.

• Recognizing that mathematics is a dynamic and changing field, not, as it is generally taught, a static and bounded discipline reflecting recorded knowledge (Confrey, 1985). Three current trends have deep implications for what it means to be literate in mathematics (Hilton, 1986): the increasing variety of applications in many other fields, which need to be recognized and understood at some level by nonmathematicians; a new unification of mathematics, which calls for breaking down artificial barriers between topics in a student's education; and the changes that the computer is bringing about in mathematics (the relative importance of topics, how some mathematics is done, and the creation of new topics), which need to infuse mathematical knowledge and understanding at all levels.

Practical Literacy in Mathematics Practical literacy is knowledge that can be put to immediate use in improving basic living standards. The ability to compare loans, to figure unit prices, to manipulate household measurements, and to estimate the effects of various rates of inflation brings immediate real benefit. This kind of applied arithmetic is one objective of universal primary education.

Civic Literacy in Mathematics Civic literacy involves more sophisticated concepts, which enhance public understanding of legislative issues. Major public debates on nuclear deterrence and nuclear power, economic policy, public health, and the use of resources frequently center on scientific issues. Inferences drawn from data, projections concerning future behavior, and interactions among variables in complex systems involve issues with essentially mathematical content. A public afraid or unable to reason with figures is unable to discriminate between rational and reckless claims in the technological arena. Ideally, secondary education should provide all students with the mathematical knowledge and understanding needed by today's "enlightened citizenry" that Thomas Jefferson called the only proper foundation for democracy.

Using Mathematics as a Tool Literacy that involves using mathematics as a tool encompasses the mathematics necessary to study and work in science, engineeering, and other fields that employ mathematical language, ideas, and models. It refers to all uses of mathematics—whether in theoretical physics or business management. As science and industry come to depend increasingly on mathematical tools, professionals in ever more diverse fields will need to learn this universal language. The basis for the mathematics that constitutes use-related literacy must be laid at the secondary school level, even though these tools are greatly extended and enhanced in college mathematics courses.

Cultural Literacy in Mathematics Cultural literacy in mathematics, the most sophisticated of these levels, pertains to the role of mathematics as a major intellectual achievement. Because cultural literacy lacks an immediate, practical purpose, its appeal may be limited. Yet the simpler and historically earlier parts of mathematical invention, like the invention of zero or of negative numbers, are accessible to many people, including quite young students. At this level of difficulty, an appreciation of mathematics as an intellectual activity engaged in by one's fellows should be part of any concept of mathematical literacy. As one progresses through the more complex developments in mathematics, however, the size of the interested audience may decrease, to an audience perhaps something like the readership of *Scientific American*. Pursuing cultural literacy in mathematics to the more advanced stages enables one to appreciate the seemingly arcane research of twentieth-century mathematics not only for its potential and unknown practical application but also, and more important, as an invaluable and profound contribution to the heritage of human culture. For the most part, individuals attain this sort of literacy through intensive study in some advanced subject, not necessarily mathematics itself.

A CONCEPTION OF SCHOOLING

How do schools produce the learning that is entailed in scientific and mathematical literacy? A central principle that guides this report is that teachers and students are the most important resources in the educational process and that their behaviors determine schooling outcomes. A second, related principle is that incentives and constraints influence the behavior of students and teachers. A third principle concerns the question "Excellence for whom?" The committee is concerned not only with the achievement levels of the most able students, but also with the distribution of knowledge and skills among students from different backgrounds. We expand our conception of these three principles in the sections that follow.

Schooling as the Behavior of Students and Teachers

The committee's formulation of indicators is based on the view that what students and teachers do determines how much learning takes place. This principle may seem obvious and not worth emphasizing. To appreciate its significance, it is useful to review how it evolved from earlier work on the determinants of children's academic achievement. Such work comes from several different disciplinary approaches—psychology, sociology, and economics.

In educational psychology, there is a long history of research on how students learn and how teachers teach. Research on learning dates back to the behaviorist theories of Thorndike (1932) and Skinner (1953, 1968), was followed by theories that emphasized the interaction of the student with the structure of the subject matter (Brownell, 1947; Piaget, 1954; Bruner, 1960, 1966; Gagné, 1965; Ausubel, 1968; Dienes and Golding, 1971), and is currently developing into theories of how children actively construct knowledge for themselves through their interaction with the environment, including the formal and informal teaching to which they are exposed (Resnick, 1987). Each of these theories has implications for the behavior of teachers as they shape their instruction.

Sociology and anthropology also have contributed insights on the effects of teachers' (and administrators') behavior as they set the context for learning by the way classroom lessons are presented, children are grouped within the classroom for instruction, and classrooms and schools are organized. (For a review, see Committee on Research in Mathematics, Science, and Technology Education, 1985:26-34.)

The 1960s saw the application of economics to the study of education, sometimes referred to as the estimation of educational production function models. The goals of this line of research, as exemplified by the widely known report by Coleman et al. (1966) on equality of educational opportunity, is to find schooling inputs that are systematically related to student learning. Initially, this research treated in parallel fashion such inputs as physical facilities, teaching materials, and the attributes of teachers and students.

These different streams of research in education have provided a great deal of knowledge about the kinds of variables that are important in explaining student achievement, including the finding that the most important resources in the educational process are human beings, whose behavior influences what is learned in school. The aspects of human behavior that influence students' achievement are wide-ranging: they include the decisions of talented college graduates about whether to become teachers and how long to stay in teaching (Schlechty and Vance, 1983), the decisions of elementary school teachers about how much time to allocate to mathematics and science (Weiss, 1978), and the decisions of students about whether to take science and mathematics courses (Welch et al., 1982; Bryk et al., 1984) and how much homework to do or how much television to watch (Walberg et al., 1986).

Although the results of educational research studies regarding the critical importance of the behavior of students and teachers have been informative, it has been difficult to make linkages between these results and policies to improve schooling. One reason is that, as these very studies indicate, the resources most important in explaining children's achievement are the human beings whose behavior influences what is learned in school. And human behavior is not subject to easy adjustment by managers and policy makers who wish to improve learning. Policy makers can change the behavior of teachers and students only to a limited degree.

Incentives and Constraints

In emphasizing that the behavior of students and teachers is difficult to alter, we do not mean to imply that it cannot be influenced. In fact, a second principle underlying the recommendations in this report is that the behavior of teachers and students is indeed influenced by the incentives and constraints they face. Examples of such incentives include teachers' salaries relative to those offered in other professions, which may attract or dicourage talented individuals, and the quality of the mathematics and science courses available in a school, which may increase or decrease student enrollment.

These two principles—the importance of the behavior of teachers and students and the responsiveness of the behavior of teachers and students to the incentives and constraints they face—have influenced both the design of this report and our recommendations. They have led us to recommend the collection of information on many aspects of the behavior of teachers and students that influence the quality of mathematics and science instruction and that ultimately influence the level of science and mathematics literacy in the population. And they have led us to recommend the collection of information on many incentives and constraints that influence the behavior of teachers and students.

From this perspective, what is the importance of physical resources devoted to mathematics and science instruction, such as laboratories, teaching materials, and, most important, curriculum? Don't they matter? Indeed they do. However, we believe that they matter primarily through their influences on the behavior of teachers and students. For example, the lack of adequate laboratory facilities may make it difficult for a school to attract teachers who really want to teach science and may force teachers who do teach science in that school to base instruction on memorizing facts rather than on developing an understanding of scientific principles through hands-on experiments. By the same token, the lack of facilities and the consequent dullness of the instruction may lead students to avoid taking science courses.

Our emphasis on looking at physical facilities and curriculum from the perspective of examining how they influence the behavior of teachers and students is not intended to downplay the importance of these resources. The opposite is in fact the case. Some of the early production function research concluded that physical facilities do not matter, because the research was based on a design that implicitly held constant which teachers worked in a school and which courses students took. This research design eliminated some of the most important mechanisms through which facilities *do* matter: by influencing the quality of teachers who are attracted to the school and the number of students who take science courses. (For a discussion of research on the effects of instructional resources, see Carey, 1986.) Thus, our emphasis is intended to highlight the potential importance of facilities in influencing the behavior of the key actors in the educational process.

Similarly, understanding the effects of curriculum on student learning is often clouded by the lack of distinction between the curriculum laid out in state and school district manuals, what has been called the mandated or intended curriculum, and the curriculum that children actually experience, the de facto or actual curriculum. The difference between the intended curriculum and the actual curriculum stems from the decisions teachers make about what aspects of the intended curriculum to emphasize and how to adapt the curriculum (including the textbook) to accommodate their own skills and interests and their perceptions of their students' skills and interests. As a result of these decisions by individual teachers about how to use the intended curriculum, children in different classrooms and in different schools experience different actual curricula and consequently learn different things, even when they all attend schools using the same intended curriculum. For this reason, the recommendations presented in Chapter 7 on indicators of curriculum quality are sensitive to the distinction between intended curricula and actual curricula.

Another implication of our perspective is that it is important to pay attention not only to the quality of the physical resources and curricula in schools, but also to the role teachers play in shaping curricula and in deciding what supplies and materials are purchased. For example, teachers are much more likely to use new curricula and new teaching materials if they have had a hand in the planning and decision processes (Berman and McLaughlin, 1974–1975). Therefore, some of our recommendations include ideas for learning more about what influences teachers' responses to changes in resources and the intended curriculum.

The Distribution of Excellence

A third principle underlying the recommendations in this report is that, in addition to describing the extent to which schools are making progress in promoting excellent mathematics and science education, indicators should address the question: Excellence for whom? This is central to promoting scientific and mathematical literacy for all students and to ensuring that talent will be nurtured wherever it is found. An example of the committee's concern regards teacher qualifications: one needs to know not only about changes in the qualifications of the nation's science teachers as a whole, but also about the qualifications of science teachers who teach identifiable groups of children, such as minority group children, urban children, rural children, and children not in advanced-placement science courses. This principle underlies many of our specific recommendations for how data should be collected and reported, especially data on teacher qualifications and on student behavior.

3
What Are Indicators?

DEFINING INDICATORS

Identifying the domains that need to be monitored is the first step in developing indicators of the quality of science and mathematics education. The next step is to define what indicators are and how they should be distinguished from such other data as simple descriptive statistics or various kinds of qualitative information. In its earlier report (Raizen and Jones, 1985:27–28), the committee defined an indicator as "a measure that conveys a general impression of the state or nature of the structure or system being examined. While it is not necessarily a precise statement, it gives sufficient indication of a condition concerning the system of interest to be of use in formulating policy." For a statistic or measure to be used as an indicator, it must have a reference point so that a judgment can be made whether the condition being described is getting better or worse (Oakes, 1986). The notion of judgment has been integral to the development of social indicators, as reflected in an early report by the U.S. Department of Health, Education, and Welfare (1969:971):

> [An indicator is a] statistic of direct normative interest which facilitates concise, comprehensive and balanced judgement about the condition of major aspects of society. It is, in all cases, a direct measure of welfare and is subject to the interpretation that if it changes in the "right" direction, while other things remain equal, things have gotten better, or people are better off.

The literature on indicators is huge (White, 1983), and so an exhaustive treatment here of distinctions between indicators and other types of information is impractical. But a recurring theme that runs through much of this literature is that indicators usually imply a causal theory or model of how some underlying process operates to generate a particular value of the indicator. This distinction is evident in the following definition (Carley, 1981:67–68):

> Social indicators, virtually by definition, specify causal linkages or connections between observable aspects of social phenomena, which indicate, and other unobservable aspects or concepts, which are indicated. This can only be accomplished by postulating, implicitly or explicitly, some causal model or theory of social behavior which serves to relate formally the variables under consideration. All social indicator research represents, therefore, some social theory or model, however simplistic. Much research to date laying claim to the term "social indicator" research consists either of descriptive social statistics, which some have argued are not social indicators at all, or of implicit postulations of causal linkages.

To be sure, all indicators are in some sense statistics, although the reverse is not so clear. Figures on crime rates are obviously important social indicators, but are the "number of police officers per capita" social indicators as well? Yes and no. They may be indicators of the value a society places on security, they may indicate the presence of an oppressive regime, they may indicate the extent of patronage, and they may also indicate crime rates indirectly. The point is that the theory connecting "number of police officers" to some condition in society is considerably more tenuous and remote than "number of murders" or "number of property thefts" per capita.

The same logic applies to changes in an indicator versus changes in a statistic. There is virtually universal agreement on the right direction of a change in crime rates, but the right direction of a change in number of police officers (or any other group for that matter) is open to debate.

How should indicators be used in policy formulation? To answer this question requires knowledge about the goals of a society as well as a theory about the nexus of causal linkages and processes that combine to produce the indicator. An unfortunate limitation of all indicators is that, while they can inform about the state of their respective domains, they cannot tell how the observed changes have come about. They cannot tell what, precisely, to do about the situation. Once the choice has been made on what social condition to

WHAT ARE INDICATORS?

assess, indicators are neutral, summary snapshots of that condition. Their implications for policy and action derive not from some inherent property they possess, but rather from the theory that the policy maker has about the underlying processes. However, it is possible to increase the utility of indicators to policy makers by ensuring that, to the extent possible, they:

- consist of reliable and valid information that is as closely related to an important aspect of the educational system as possible,
- have reasonably direct policy implications,
- be small in number, and
- be easily understood by a broad audience.

In consideration of these criteria, the committee has grouped its recommendations on indicators into three categories: (1) key indicators that are or would be feasible given adequate investment in experimentation and development and that should be included in even the most parsimonious monitoring system, (2) supplementary indicators that are presently feasible or might be developed, and (3) research on hypothesized causal links among some important but poorly understood aspects of education in order to create and validate indicators related to these aspects.

INTERPRETING INDICATORS

Once a value has been established for a given indicator, there are essentially three possible interpretations, all of which involve comparisons of some sort. First, the value of the indicator might be compared with some absolute standard. For example, professional consensus might be used to establish a "minimum knowledge level" of a new K–5 teacher. An indicator of this could be scores on a pencil-and-paper test to measure the amount of knowledge attained by teachers. Interpretation would involve comparison of the teachers' scores with the absolute standard. (It should be noted in passing that absolute or ideal values for most indicators are difficult to establish.)

A second interpretation involves comparison of a given indicator value with its value at some prior time. For example, the percentage of high school students who took a physics course in a given year might be compared with the percentage who took a physics course in some prior year.

Third, indicators can be presented as a basis for the comparison of instructional programs, demographic groups, states, regions, countries, and so on. The proper interpretation of such comparisons is limited because of differences in social, political, economic, cultural, and other characteristics. Nevertheless, when data are disaggregated on any basis and presented side by side on a page, the temptation to make evaluative comparisons, whether warranted or not, is overwhelming and nearly universally succumbed to.

Problems in interpreting educational indicators fall into three broad categories and are sufficiently pervasive to merit brief mention here, together with suggestions for avoiding or at least minimizing their adverse consequences. The problems are (1) choice of variables, (2) levels of aggregation, and (3) scale. These problems of interpretation have to be faced before data collection can begin.

Choice of Variables

Even after the key domains to be monitored have been identified—for our purpose, student learning, general scientific and mathematical literacy, student behavior, teaching quality, curriculum quality, and financial and leadership support—the number of possible variables from which to choose in constructing indicators of science and mathematics education remains large; a partial list could well number over 100. According to the committee's formulation, various teacher and student behaviors and the incentives and constraints that influence them are presumed to be causally related; for example, the quality of the curriculum and the use of it made by the teacher affect student competence in science and mathematics and student attitudes. To what extent will the conclusions one draws from one combination of variables be similar to the conclusions one would have drawn had a different set of variables of the same underlying condition been used to construct the indicator? The answer to this hypothetical question depends critically on the quality of the sets of variables and the manner in which they were combined. One gets an entirely different picture of the educational health of the nation depending on whether one looks at high school dropout rates, results from the National Assessment of Educational Progress (NAEP), student career choices, or amount of homework assigned per pupil. Each variable or combination of variables highlights a different aspect of the complex construct "educational health." The accuracy and appropriateness of interpretations and policy decisions are limited by the quality of the indicators themselves and the manner in which they are combined.

Problems of Aggregation

When data are aggregated from one level (e.g., students) to another (e.g., classrooms, schools, or districts), numerous interpretive difficulties arise. Data should be collected and aggregated according to a clear conception of schooling and with a view of who will use the information and for what purpose. Data aggregated to levels that are inappropriate to relevant policy decisions may be quite misleading—for example, statewide averages on teacher salaries may not be useful information for a particular school district. In general, data at the level of the individual student are most useful to that student's teacher, classroom-level data are of most interest to principals, school-level data are most useful to superintendents, and so on.

Aggregation Effects and the Ecological Fallacy Levels of aggregation exert important effects on correlation coefficients. These effects help to explain why the results of educational research vary so much from study to study. What, for example, is the correlation between socioeconomic background and achievement test scores? Is it .3? .6? .9? All three are possible. The correlation depends on the unit of analysis, the population sampled, and the way the two constructs are measured. If one takes these three factors into account, the results are fairly consistent.

Using national samples of high school students, family income correlates about .3 with achievement test results at the student level. Aggregating to the school level, the correlation is between .5 and .6 among school means nationally. If, however, one looks within large urban districts, the school-level relationship is between .8 and .9. The district-level relationship varies from state to state (.2 to .6), and at the state level the correlation between 1975 poverty rates and state achievement estimates is .63 (N = 50 states). Table 3-1 summarizes these results.

Other differences are found when looking at different grade levels, or when indicators other than poverty are used to represent home background. For example, in Project TALENT, an indicator of socioeconomic environment based on home variables that were hypothesized to exert a more direct effect on achievement (mother's education, books in the home, child has own desk, etc.) correlated .5 at the student level for high school students (Flanagan and Cooley, 1966).

TABLE 3-1 Socioeconomic Background and Achievement

Level	Population Sampled	SES Indicators	Correlation
Student	National	Income	.2 to .4
Student	National	Home environment	.5
School	Large urban district	Income	.8 to .9
School	National	Income	.5 to .6
District	Within state	Income	.2 to .6
State	National	Income	.6

Source: Cooley et al. (1981).

What is the appropriate unit of analysis? It depends, of course, on the question being asked. A scatterplot depicting the modest relationship between socioeconomic status (SES) and achievement at the student level is typically an oval-shaped swarm of points with few outliers. Given this fact, inferring from the within-district school-level correlation of .9 that most low-achieving students come from poor homes is an excellent example of what sociologists call the ecological fallacy: the error of using relationships at one level, such as school, to describe relationships at a lower level, such as student (Robinson, 1950).

Correlations at one level of analysis differ from correlations at another because of the grouping effect. This occurs when membership in the group (e.g., class or school) is related to either one or both of the variables being correlated. For example, the socioeconomic homogeneity of neighborhoods produces a relationship between SES and school, and that relationship produces the larger correlation between SES and achievement at the school level than at the student level.

Many statisticians would argue that the proper procedure is not to use correlations at all when, as in the case illustrated, regressions are appropriate (see, e.g., Cain and Watts, 1970). However, the use of correlations is so universal in analyzing and reporting educational data that we consider it important to warn against misinterpretations. Our brief discussion of problems in "ecological inference" merely scratches the surface. A detailed and comprehensive (although not too technical) treatment is provided by Langbein and Litchtman (1978).

Inconsistent Aggregation and Self-Selection Every student of elementary statistics is warned early in instruction that teasing out causal relations among any set of variables can be a tricky and often misleading endeavor. It is surprising how often unwarranted causal conclusions are drawn from summary indicators, whether or not the persons involved have had training in data interpretation. The temptation, for example, to judge the quality of education in a state by the mean Scholastic Aptitude Test (SAT) scores of its graduates, despite cautionary statements issued by the College Board (e.g., Hanford, 1986), is a case in point. This annual practice illustrates in a nutshell most of the pitfalls considered in this section.

Why are mean SAT scores inappropriate indicators of the comparative quality of instruction in the various states? First, consider the problem of sample representativeness. How representative are students who take the SAT of the typical high school graduate? In general, college-bound seniors (the SAT population) are better prepared academically than their noncollege-bound counterparts. Moreover, there is wide variation in the percentage of students by state who take the SAT. (Some state institutions of higher education require SAT scores for admission; some do not; others require scores on tests administered by the American College Testing Program.) For example, in 1984, the percentage of high school seniors by state who took the SAT ranged from a low of 3 to over 65 percent. For various reasons, including self-selection, the smaller the percentage of students taking the SAT, the higher their mean SAT scores. Thus, inconsistent aggregation leads to false and misleading comparisons.

Problems of Scale

The first interpretive problem in this category involves the actual scale itself. Should absolute values (number of science and mathematics teachers, number of students taking at least two years of mathematics, etc.) be used, or should various ratios (for example, science teachers per 100 or 1,000 pupils or the ratio of science and mathematics teachers to all teachers) be used? Often—but not always—ratios and proportions are more informative reporting units. A simple example illustrates why this is so. An absolute increase in the number of unemployed persons who are actively seeking employment is generally agreed to be a move in the wrong direction. But such an increase, by itself, may be misleading. If the entire labor force has increased significantly, it is possible that an increase in the absolute number of unemployed actually represents a decrease in the unemployment rate, that is, the percentage of the labor force that is unemployed. A counter example from education involves increasing the length of the elementary school day and introducing an additional subject, say, health and family education. Under these circumstances, the proportion of school time devoted to mathematics might decrease, an apparent move in the wrong direction, but the number of minutes per day given to mathematics might actually increase.

In many situations, it is wise to collect information on and report both types of figures. For example, it may be important to know both the absolute number of minutes per day a student spends doing mathematics as well as the percentage this figure represents of the student's total time spent on school work.

Another scale issue that, surprisingly, often goes overlooked is the use of scale units that change over time or that have different meanings in different locations. The most commonly used units are those involving monetary values. Total school budgets, dollars spent on laboratory equipment, and teacher salaries are all examples of scale units that vary to the extent that the value of the dollar varies over time and over locations. Results not adjusted for this variation may seriously distort the picture. Thus, total school expenditure should be adjusted to total expenditure per pupil, with perhaps an additional adjustment for variations in the cost of living; teacher salary should similarly be adjusted to account for local cost of living, and so on.

INDICATORS FOR WHOM?

This chapter argues that an indicator is more than another layer on a mound of statistics; rather, it can be used in a systematic attempt to investigate the interaction among selected pieces of information. Federal, state, and local education bureaucracies are awash in numbers. The challenge taken up by the committee in this report is to go beyond an endless parade of statistical tables and focus on the key questions and subsequent indicators that will be credible to policy makers in state and local education agencies—the major decision makers since education in the United States is overwhelmingly a state and local legal and fiscal responsibility. The challenge for state and local policy makers is to adopt and use the indicators that, when combined, best represent a snapshot of what exists today in mathematics and science education as well as point to promising policy initiatives.

At all levels of the education system, there is recognition of the need for a reliable and valid evaluation of how well students know, understand, appreciate, and use information they have received in their K–12 mathematics and science experience. And, as with any evaluation, the initial temptation is to start to collect data before the key questions have been asked. Once the questions are specified, most of the data can probably be obtained without generating a new national information system that may fall under its own weight (see Appendix E). In this respect, a concern shared by the committee and state administrators alike is feasibility. By feasibility, we mean that collection, analysis, and reporting of valid data should be possible in a timely manner, given reasonable resources. The design decisions and availability of resources that affect the frequency of collecting data, as well as methodology, may well be driven by timetables that allow indicators to interact with and influence policy.

COLLECTING INFORMATION

Once decisions have been made on the type of indicator to be used (e.g., student test scores, teacher salaries, judgments of curricular quality), there arises the question of how to collect the pertinent information. This report argues that a wide range of data-collection methods is necessary. Some of the recommended methods have been used extensively in the past, such as surveys; others are less widely used, such as time-use studies. The key challenge is to tailor the proposed data-collection methods to the type of information that is needed.

Comparability Versus Depth of Information

There is a difficult tension in the choice of data-collection methods between collecting comparable data and being open to unexpected responses. For example, closed-ended questionnaires produce standardized information comparable across space and time and are particularly suitable for collecting information on such matters as salaries and defined fringe benefits, for which comparability is critical, and the nature of the desired information is relatively clear-cut.

Closed-ended questionnaires are poorly suited, however, to the collection of information dealing with such topics as how teachers and students spend their time outside school. The reason is that the range of possible responses is much broader than can be captured by a closed-ended questionnaire. Consequently, it is important to give up standardization in favor of capturing diversity. Thus, time-use studies are more appropriate for collecting this type of information.

A related issue arises in attempts to improve achievement tests, questionnaires, and the like so that responses mirror more faithfully and in greater depth, say, what students have learned and are able to do. Two problems arise: first, to the extent that items, examples, and questions are improved to capture more and better information, comparability to earlier assessments is lost. Second, assessments are likely to become more costly, and sample sizes may have to be reduced. This may create loss of generalizability (as in studies using classroom observation), although matrix sampling and other techniques may partially overcome this problem. These problems are not cited to argue against improving assessment instruments and questionnaires—we argue quite the contrary in the next chapter—but only to sensitize those using indicators to some of the difficulties involved in designing the requisite collection of data and information.

Timing

How often should information be collected? There is tension between the expense of collecting information often and the value of up-to-date information that permits rapid discernment of changes in trends. The choice of how often to collect data for a particular indicator should depend on the importance of the indicator for informing policy and on how rapidly changes are likely to occur in the distribution of the behavior, incentive, or outcome reflected in the indicator. Consequently, we argue for the assessment of student learning at given grade levels every four years, except for science achievement in elementary school, for which the current improvement efforts warrant assessment every two years. No matter what the frequency, it is important that each wave of information be collected at the same time of the year so as to maintain consistency and provide comparable data.

Design of Expert Panels for Assessment

At various places in this report, the committee recommends the use of panels of experts as a method for assessing instructional materials and performance when no suitable outcome measure is yet available. Because the use of experts is an often-used mechanism, we discuss the problems inherent in its application in some detail.

Based in part on our experience with difficulties encountered in the experiment on reviewing the science content of science achievement tests (see Appendix B), we consider it important to make some general comments about the use of expert panels as an assessment method. First, there should be a clear understanding among the panel members as to the intent and interpretation of the material to be judged or rated. Second, if the tests or other materials are to be used for various purposes, the panel members should understand and the ratings should distinguish among these purposes. Third, there should be agreement as to the rating criteria. Panels can meet these three conditions by using rater "training" exercises or discussing their procedures before the actual work begins. Discussion of the ratings by panelists after they have completed their work may further help to clarify whether purposes of the materials and rating criteria were unambiguous. (However, it is not desirable that the panel members change their ratings as a result of the post-rating discussion, at the risk of reducing independence of the panelists' ratings.) Such techniques help to improve the rating process and to reduce the variability between raters.

Rater Variability Variability between raters with regard to individual items is one source of variability in panel assessments. However, the scores of an individual rater on different items tend to be correlated. This correlation is one quantification of frequently heard comments, such as that one rater tends to give high scores and another low scores. It is not generally recognized that, as a result, the impact of rater variability on the variability of average scores or percentiles can be substantially greater than indicated by the variability between raters item-by-item, perhaps by an order of magnitude. In the experimental review of science achievement tests, this was true not only of types of reviewers (teachers, scientists) but also of reviewers within type. It is not feasible to eliminate these sources of rater variability. Thus, panel studies should be designed to provide estimates of rater variability and correlated variability. Such information has the potential for improving the design of expert panels, for example, for deciding on the number of panel members needed to yield acceptably reliable estimates of averages, percentiles, or other statistics of interest. With a positive correlation between the ratings of an individual reviewer by item, the use of a given number of reviewers, each rating every item, will yield less reliable statistics than a larger number, each rating a randomly chosen subsample of the items. This may be potentially useful when there is a large number of items to be rated and the rating process is time-consuming. Appreciation of the sources of rater variability will also help ensure that standard errors of statistics derived from panel ratings are properly computed.

Validity and Reliability The design of an expert panel should consider the problems of both accuracy (validity) and precision (reliability). The concept of accuracy implies that there is a "true" value to be estimated. The true value may have a theoretical definition or may be defined only operationally as that value resulting from a set of carefully specified empirical measurement steps. A panel whose assessments differ systematically, in either a positive or negative direction, from the true values is "biased." In experiments such as the science test review, the standards against which raters assign their scores are critical since they affect the accuracy of the scores as measures of the relative value of alternative tests. Depending on their biases, reviewers may give a poor test relatively high ratings and a good test relatively low ratings so that two tests that differ widely in their true value are judged—on the basis of average ratings—to be equally effective. Similarly, ratings of teacher performance based on classroom observation are likely to be strongly affected by the personal views of the observer regardless of the procedures established for the assessment. The steps outlined above will help to minimize biases due to misunderstandings on the part of panel members. They will also improve the interpretation of the ratings. It may be possible to design a questionnaire for potential panel members that would help ensure ratings free of personal preference or provide a basis for eliminating the ratings of particular individuals.

Coordination of Strategies for Collecting Data

In each of the chapters that follow, recommendations are made for data to be collected or observations to be carried out or both. Implementation of these recommendations will involve surveys and other data collection strategies that should be coordinated. It is not the committee's intention that whole, new data systems be set up to carry out its recommendations. Instead, several existing mechanisms currently undergoing review and reformulation should be used to implement the recommended data collections and analyses, including the redesigned elementary/secondary data collection of the Center for Education Statistics, the Assessment Center of the Council of Chief State School Officers, and the educational data improvement effort intended to lead to common data collection by the states.

In Appendix E we discuss issues of coordination, pulling together recommendations from throughout the report that imply surveys, referring to ongoing efforts, and outlining suggestions for how desirable new survey efforts might be implemented. More intensive survey design planning including issues of sample size should be left to agencies—national, state, or local—that assume or are assigned responsibility for the indicators.

4

Indicators of Learning in Science and Mathematics

This chapter first appraises currently available multiple-choice tests of student achievement in order to judge their suitability as indicators of the quality of education in science and mathematics. This appraisal includes a discussion of various uses of these tests, a review of criticisms of current testing methods, and suggestions on some desirable features that should be retained. New methods of assessment are then described that would provide both quantitative and qualitative information about how students perform tasks requiring higher-order skills. The chapter continues with our recommendations regarding uses of current indicators of student learning and work needed to develop improved tests. Implications of these recommendations for state education agencies are presented, and we conclude with a discussion of possible approaches to assessing aspects of scientific literacy of the U.S. population.

AN APPRAISAL OF CURRENT TESTS OF STUDENT ACHIEVEMENT

The most direct indicators of the quality of science and mathematics education are the scores based on tests that measure what students have learned. Currently available indicators of student learning are typically obtained from standardized achievement tests made up of multiple-choice items. Before one accepts information based on such tests, it is necessary to make an appraisal of their suitability as indicators of the quality of education.

Purposes of Testing

In practice, tests are used for a wide variety of purposes. Some involve the evaluation of individual students for grading, student counseling, placement, promotion, awards, scholarships, and so on—important for educational purposes but not always well suited to the development of indicators. The use of tests most closely identified with assessing the condition of science and mathematics education is for the evaluation of learning achieved by populations of students; a related purpose that is of interest to the committee is the use of tests in improving the quality of instruction.

Evaluation of Student Learning Measures of the outcomes of education for students are critical indicators in any educational monitoring system. Hence, the testing purpose of primary concern to the committee is evaluation of student learning, particularly at national, state, and regional levels. Indicators of learning that are satisfactory for this purpose would also be useful to school districts or individual schools as a means of monitoring change in levels of accomplishment over time.

A related use of tests is to provide criterion measures to validate less direct indicators of the quality of education, for example, teaching effectiveness or the quality of the curriculum. Tests are often used for this purpose, but such use is appropriate only when the tests being employed assess important dimensions of student learning in a satisfactory manner.

Improving Instruction One reason for monitoring the condition of mathematics and science education is to be able to improve instruction. Several applications of tests can help do so: tests can contribute to raising the standards of schools as to the skills and competencies to be taught and acceptable levels of performance. They can provide diagnostic information that would enable teachers to understand the reasons for failures and provide appropriate remedial treatment. Diagnostic information would be useful in school assessment at local, district, state, or even national levels as well: better understanding of why students develop erroneous problem-solving algorithms or fail to modify childhood misconceptions of physical principles would make possible actions at higher administrative and supervisory levels aimed at improving instruction. Tests can also be used as dependent variables in experimental studies involving educational treatments or methodologies developed to improve instruction.

Test questions also can be used for teaching as practice exercises with feedback. Much practice is necessary to acquire the complex skills required for development of the automatic processing and pattern-perception skills that are essential for the performance of more advanced problem-solving tasks. Such exercises might also provide, as a by-product, information that would be useful for large-scale assessment of student learning. Still another instructional application is to improve the articulation of instruction at various transition points, for example, between elementary, middle, and high school or between introductory and advanced college courses. Tests can determine whether students actually possess the basic knowledge and skills necessary for successfully dealing with the more advanced concepts and procedures taught at the next educational level. Although such instructional uses of tests are not directly related to their use as indicators, they are as important and provide equally valid reasons for developing better tests.

Criticisms of Current Testing

In the early years of this century, the assessment of student achievement was generally based on teachers' judgments, which were in turn based on teacher-made tests, homework, and impressions of classroom performance. But after the demonstration of the efficiency of objective tests by the use of the Army Alpha tests in World War I, a revolution in testing methods began. The invention of the multiple-choice test item and the development of fast and efficient test-scoring machines (Lindquist, 1954) made possible the mass testing of students on a very large scale. Testing agencies and test publishers hastened to develop multiple-choice tests, teachers were trained to write multiple-choice items, and many colleges set up testing bureaus to assist the faculty in preparing and scoring multiple-choice examinations. Except for the teacher-made tests that many teachers still rely on for grading students, multiple-choice tests have driven out virtually all other types of examinations because of their objectivity, speed, and economy (N. Frederiksen, 1984a).

From the standpoint of assessing the quality of education in science and mathematics, it is important to know to what extent information based on tests in current use provides a sound basis for judgment. Standardized multiple-choice achievement tests have been widely criticized not only by educators but also by students, parents, and the media. Some of the criticisms most relevant to the development of indicators of science and mathematics education are discussed below.

Multiple-Choice Tests Penalize Creative Thinking This is a well-taken criticism, since most multiple-choice items do not provide much opportunity to generate new ideas. Students responding to a typical multiple-choice item begin by reading the stem (the expository or question part of the item); then they read the first option and make a narrow directed search of their memory store to find a match to the option. If they find information that clearly matches the option, they may mark it and go to the next item. If not, they read the next option and again seek a match and mark it or consider the next option, and so on until they either choose and mark an option that matches information stored in memory or skip the item. Such a process would appear to require little creative thinking. Of course, some multiple-choice items require more complex processing of information, but a large majority of the items in a typical achievement test measure factual knowledge.

In spite of the controversy, there has been little research on the mental processes involved in taking a multiple-choice test. Several investigators, however, have compared multiple-choice tests with their free-response counterparts, which were constructed by substituting an answer space for the multiple-choice options for each item (Vernon, 1962; Traub and Fisher, 1977; Ward, 1982; Webb et al., 1986). As judged by correlations and other statistical analyses, the format of the test was found to make little difference. With a few minor exceptions, for tests that were originally constructed with multiple-choice questions, both formats appeared to measure the same ability.

However, use of the multiple-choice format may tend to exclude the writing of items that require more complex thinking processes. If so, different results might be found if one began with free-response problems intended to elicit productive (rather than reproductive) thinking and converted them to the multiple-choice format. Such a comparison was carried out using a test that required students to formulate hypotheses that might account for the findings of an experiment (N. Frederiksen and Ward, 1978). Indeed, quite different results were obtained than in the conversion from multiple-choice to free-response formats. The correlations between the two formats were generally low, and the pattern of relationships to various cognitive abilities was different. The two formats were similar with regard to their relationships to verbal ability and reasoning, but only for the free-response version were there substantial relations to a factor called *ideational fluency,* which represents the skills involved in making broad searches of the memory store in order to retrieve information relevant to a situation (Ward et al., 1980). In at least one instance, converting a test intended to measure productive thinking to multiple-choice format eliminated the need to broadly search the memory store for ideas that might be relevant, evidence that the multiple-choice format is not conducive to measuring productive thinking.

Multiple-Choice Tests Are Not Representative of Real-Life Problem Situations There are at least two aspects of representativeness. One has to do with the frequency with which real-life problems occur in multiple-choice form. Occasionally people encounter problems with a limited number of clearly defined options, such as deciding whether to go left, right, or straight ahead at an intersection, or whether to take the morning or the afternoon flight to Miami. But more often there are many options, and one does not know what they are and must think of them for oneself. Multiple-choice options are almost universal in educational testing but rare in real life.

The other aspect of representativeness has to do with the extent to which the problems posed by test items are similar to problems that occur in real life. Problems encountered in real life generally involve situations that are far more complex and richer in detail than are provided by the stem of a multiple-choice item. Furthermore, there seems to be a tendency for testers to use stereotyped sets of test problems in both science and mathematics, problems that, for example, involve weights on inclined planes, pulleys, boats going with or against the current, and the number of pencils Jane has. Generalization of learning would be facilitated by schoolroom experiences that resemble problems in the world outside the classroom with respect to the variety and complexity of problem situations and settings. Use of test problems that simulate such situations would encourage such instruction (Norman et al., 1985).

Multiple-Choice Tests Are Undesirably Coachable Any test is coachable in some sense and to some degree. Some kinds of coaching involve training that has nothing to do with the subject matter of the test, such as teaching students that the longest multiple-choice option is most likely to be correct and to avoid highly technical options; in such cases coaching may improve test scores somewhat without improving the ability presumably measured by the test. Another kind of coaching attempts to improve the ability measured by the test; a review of fractions and percentages, for example, might improve both test scores and the student's underlying competence in arithmetic. Test makers should attempt to construct tests that are coachable only in the sense that coaching and teaching are indistinguishable. Tests that are coachable in the undesirable sense not only result in wasted time; they also tend to falsify the data.

It is difficult to estimate the size of gains that are attributable to coaching (Messick, 1980). Most coaching is probably done by teachers in school settings and generally consists of attempts to teach the kinds of knowledge and skills that are measured by the tests. Coaching schools are more likely to attempt to teach test-taking skills, with less attention to the content of the test; fantastic gains have been claimed for such coaching (Owen, 1985), but without much evidence. The studies of coaching for the Scholastic Aptitude Test (SAT) and similar tests that were reviewed by Messick show modest gains on the average—less than 10 points on the SAT-verbal and about 15 points on the SAT-mathematics test, on a scale of 200 to 800. The gains are difficult to interpret, however, because of variations in methods of assigning students to the coached and control groups (often the coached students are volunteers), the methods, length, and content of coaching, and methods of analyzing the data. Thus, it is usually difficult to judge whether gains are attributable to (a) differences in ability or motivation, (b) the nature and length of the coaching, or (c) the methods and variables used in attempting to control statistically for differences between the coached and control groups. Messick suggests that the smaller effects seem to be associated with short-term cramming and drill and the larger effects with longer-term programs involving skill development—especially in mathematics, for which there is likely to have been greater variability with regard to opportunities or motivation for students to learn.

Such results suggest that coaching is not likely to produce major distortions in the distributions of scores obtained from current tests.

However, even small average gains could lead to mistaken conclusions when test scores are used to monitor change in student achievement.

Multiple-Choice Tests Exert Undesirable Influence on the Curriculum There are many reasons to believe that the nature of the tests used influences what teachers teach and what students learn. Students want to get respectable grades, or at least pass the course, and teachers believe that they may be evaluated on the basis of their students' test scores. Tests that fail to match the intended curriculum may therefore have undesirable effects on learning.

Testing had relatively little impact on instruction in the 1950s and early 1960s, but the situation began to change in 1965 when the Elementary and Secondary Education Act (ESEA) was passed. The act required that certain teaching programs funded by ESEA be evaluated, and future funding of programs often depended on the outcomes of the evaluations (Popham, 1983). Pressure to improve test performance increased during the 1970s, when test data showed that attainment of knowledge and skills was declining (Womer, 1981), and the National Assessment of Educational Progress (1982) reported decrements in performance. Still more pressure to "teach for the tests" resulted from the decision of a federal judge in 1979 that Florida's use of a competency test to satisfy graduation requirements was unconstitutional unless preparation for the test was provided.

Educators representing a majority of the school districts identified by the National Science Teachers Association as exemplary in the teaching of K-6 science (Penick, 1983) have expressed concern at the mismatch between currently available standardized tests and their curricula. These districts are teaching inquiry-based, hands-on science, which both the scientific and educational communities strongly support, but the skills acquired by their students are not measured by the tests. At a conference on elementary science education held by the National Science Resources Center at the National Academy of Sciences in 1986, participants representing school districts with innovative programs expressed concern "that standardized achievement tests do not do a good job of assessing what students learn in elementary school science. There is a need to develop improved tests and alternative evaluation techniques to assess student progress in science, with more emphasis on the development of process skills and attitudes" (National Science Resources Center, 1986:3). As more school districts are striving to introduce more effective science programs in grades 1–6, the issue of correspondence between tests and curricular goals becomes particularly critical at this level.

Bloom (1984) wrote that "teacher-made tests (and standardized tests) are largely tests of remembered information. . . . It is estimated that over 90 percent of test questions the U.S. public school students are now expected to answer deal with little more than information. Our instructional material, our classroom teaching methods, and our testing methods rarely rise above the lowest category of the [Bloom] taxonomy—knowledge" (p. 13). Resnick and Resnick (1985:15), in commenting on state testing programs, stated

> It is appropriate . . . to think of minimum competency programs as an effort to educationally enfranchise the least able segment of the school population. . . . However, by focusing only on minimal performance, the competency testing movement has severely limited its potential for upgrading education standards. Only recently have some states begun to include higher level skills in their competency testing programs. It would be difficult to stress too much the importance of this move beyond the minimum . . . for there is evidence that examinations focused solely on low level competencies restrict the range of what teachers attend to in instruction and thus *lower* the standard of education for all but the weakest students.

An examination of the results of state testing programs in mathematics provides further documentation: children score well on items dealing with computation but less well on items dealing with concepts and problem solving, because the learning of these higher-order skills is not stressed in classroom instruction (Suydam, 1984).

The National Assessment of Educational Progress (NAEP) report (1982) previously referred to showed similar results. Performance by comparable populations of students on test items measuring basic skills did not decline compared with earlier assessments, but there was a decrease on items reflecting more complex cognitive skills. In mathematics, about 90 percent of the 17-year-olds could handle simple addition and subtraction, but performance levels on problems requiring understanding of mathematical principles dropped during the preceding decade from 62 to 58 percent. In science, performance declined for both kinds of items, the decrease being twice as large for items requiring more advanced skills.

It seems a reasonable conjecture that the mandated use of minimum-competency tests and concurrent emphasis on basic skills was at least in part responsible for these declines. It is possible, however, to use the influence of tests on what is taught to improve learning by constructing tests that require the more advanced skills. Such tests would thus provide incentives for improving the quality of education in science and mathematics (N. Frederiksen, 1984a).

In Chapter 7, the committee recommends that basic curriculum frameworks be developed for nationwide use, frameworks that represent the best opinions of working scientists and mathematicians, as well as educators, as to what should be taught and tested—a core of essential factual knowledge and the algorithmic and procedural skills and higher-order competencies for doing real science and mathematics. Tests that match such frameworks would influence teaching and learning in desirable directions.

Multiple-Choice Tests Are Not Based on Theory This criticism is not one that is frequently voiced by critics, but it deserves mention. In one sense, multiple-choice testing is indeed based on a theory, namely, a very extensive theory of the mathematical and statistical considerations having to do with test reliability, validity, error of measurement, methods of item analysis, item parameters, equating of tests, latent trait models, and so on (e.g., Gulliksen, 1950; Rasch, 1960; Lord and Novick, 1968; Lord, 1980). This test theory is largely based on the assumption that items are scored objectively as either right or wrong, and the test score is the number right. Item-response theory, a relatively new and very influential part of test theory, assumes a multiple-choice format by taking account of guessing. This body of work has been extremely useful and important in the development of assessment methods. But none of this test theory is concerned with the content of the test items.

Another kind of theory, one that grows out of work in cognitive psychology and artificial intelligence, does provide a potentially useful basis for the development of tests based both on content and the cognitive processes that are involved in doing science and mathematics. Some of the implications of this work are described later in this chapter.

Science Content in Multiple-Choice Achievement Tests is Questionable In order to obtain information on the quality of the science content in currently used achievement tests, the committee asked 12 scientists and science teachers from several science fields to evaluate the items from 9 commonly used multiple-choice achievement tests. (Two individuals did not review the items but wrote general comments.) This attempt to evaluate tests is described in more detail in Appendix B. Since differences in average ratings between the tests were relatively small compared with the variability between the reviewers, no quantitative conclusions concerning their relative merits could be justified from their evaluations. There was agreement, however, that the tests were poor at probing higher-order skills and that they contained a significant (5 to 10 percent) number of flawed items. The remaining items were judged to be quite variable in their quality, such that it was not obvious that a positive change in test score would in fact mirror improvement in the quality of student learning. The committee's experience with this experiment in assessing science tests reinforces concern about the quality of the subject-matter content of some of the tests in common use, even while it emphasizes some of the difficulties in obtaining reliable evidence on this important question.

Some Virtues of the Current Testing System

Despite the criticisms that have been leveled by the committee and others at the current system of educational testing, it has a number of virtues that should be acknowledged. First, the multiple-choice format for testing makes possible the economical measurement of factual knowledge. This format allows the rapid and reliable scoring of tests at a relatively low cost. Therefore, it seems sensible to retain the conventional test format for doing what it does best—measuring factual knowledge and the ability to use the simpler kinds of procedural knowledge, such as the algorithms used in arithmetic computations (to the extent that they continue to be taught).

Two other useful developments in current testing systems are matrix sampling and the application of statistical methods to make possible test comparisons over time. Neither of these is limited to tests in the multiple-choice format. The use of matrix sampling allows one to obtain information about large populations of students without concomitant increases in cost and testing burden. Matrix sampling is analogous to the methods used in public-opinion polling, in that it requires drawing random or representative samples of subjects. But in addition to drawing random samples of subjects, matrix sampling also involves independently drawing random samples of test items (Wilks, 1962; Lord and Novick, 1968); thus random subsamples of students are given different subsamples of items. An adaptation of the item-sampling procedure used by NAEP involves what is called a *balanced incomplete block design* (Messick et al., 1983). This procedure makes possible the calculation of close approximations to the means, standard deviations, intercorrelations of tests and test items, and so on, that would be obtained if the entire school population had been tested. This is an important feature of the methods currently employed by NAEP. When tests are created that are more costly to administer and score than conventional multiple-choice tests, the use of matrix sampling will be critical for keeping costs within bounds.

Another virtue to note is that current testing methodology makes possible comparisons over time. The collection of data on learning indicators is of limited value unless the measurement can be repeated, since the purpose of school evaluation is to detect change—to see if student performance is improving. Given that test-score scales are arbitrary, measures taken on a single occasion may be of limited value. The only way in which such measures would be interpretable would be for the scores to have intrinsic meaning apart from comparative interpretations.

School evaluation is concerned not only with measuring change in the same individuals over a period of time but also with comparing the performance of successive groups of students at a particular stage of instruction, such as the end of the eighth grade. The latter kind of comparison is of particular interest at state and national levels. Unfortunately, it poses a difficult problem of interpretation because of possible changes in the composition of the groups that have nothing to do with instruction. And there are many other problems of interpretation due to the use of fallible instruments, the possibility (if not likelihood) that a given test does not measure the same abilities before and after a period of training, the lack of random assignment of students, the lack of equal units on a score scale, the unreliability of difference scores, and so on. (see Harris, 1963). But statistical test theory has provided workable answers to many of these problems, for example, in the development of methods of equating scores on different versions of a test (Angoff, 1984). The development of item-response theory (Lord, 1980) provides workable solutions to other problems. The extensive test theory that has been developed should be retained, but it needs to be adapted as necessary for use with new testing procedures.

NEW METHODS OF ASSESSMENT

The procedures suggested in this section, if properly developed, could provide remedies for the problems described in the use of multiple-choice tests. They could provide both quantitative and qualitative information descriptive of how students perform the most important higher-order science and mathematics tasks. The results could reflect such attributes of performance as speed of responding, use of inference in problem solving, pattern-recognition skills, students' internal models of problems, and use of strategies and heuristics in solving problems.

Two major kinds of assessment procedures are considered. One consists of what might be called *global measures,* since the performance to be elicited will be evaluated as a whole. The other set of procedures yields *processing measures,* since they are descriptive of the information-processing components that influence the development of conceptual knowledge and overt performance of the student.

Global Assessment

A frequently used alternative to a multiple-choice test is an essay test in which the items elicit fairly long written responses. Such tests have the virtue that students not only must think of the ideas for themselves but also must organize them in an appropriate sequence and state them clearly. Essay tests have been justifiably criticized, however, on the basis of the subjectivity and unreliability of scoring. Reliability can be improved by pooling the grades of two or more readers; in the case of essays written to test English-language proficiency, a holistic method of grading is used in large-scale testing in which two or more judges are asked to read each essay quickly and rate it impressionistically, and the ratings are pooled. The result is that grades are more reliable, but no one knows precisely what they mean.

Another approach that has been tried involves the use of tasks that impose more structure on the response than does the typical essay question, so that one can know more precisely what skill is being measured (N. Frederiksen and Ward, 1978; Ward et al., 1980). In science, for example, the test problems might simulate tasks that are frequently encountered by scientists, such as formulating hypotheses that might account for a set of research findings, making critical comments on a research proposal, or suggesting solutions to a methodological problem. For example, in one exercise students were asked to write down the hypotheses that they thought should be considered in trying to account for the findings (shown in a graph or table) of an experiment or field study. Development of materials to aid in scoring this kind of test requires a protocol-analysis procedure that includes the following steps: (a) making a classification of the ideas written by a sample of students, (b) writing definitions of the categories, and (c) having experts make judgments about the quality (and other attributes) of each category, in light of the information that was available to the students. Coders are trained to match each of a student's responses to a category, and scores can be generated by a computer on the basis of quality and other values attached to the categories.

Tests of this sort were found to be poorer than Graduate Record Examination (GRE) scores for predicting first-year grades in graduate school, but they were better than the GRE for predicting such student accomplishments as doing original research, designing and building laboratory equipment, and being author or coauthor of a research report. Thus, there is at least correlational evidence that tests of the kind described above measure something related to productive thinking that is not measured by conventional tests.

More sophisticated methods of analyzing free-response protocols are being developed, methods that do not require the imposition of such a high degree of structure. These methods are based on discourse analysis (C. H. Frederiksen, 1975, 1985; van Dijk and Kintsch, 1984; C. H. Frederiksen et al., 1985); they make it possible to investigate understanding by analyzing free-response productions of students. Flexible computer environments are being developed that permit students to generate text based on their retrieval, generation, and manipulation of declarative knowledge in a knowledge-rich domain. The use of syntactic and semantic parsers makes it possible to analyze a student's responses to a task and to make their grammatical structure explicit on the screen. Analysis of the structure is then possible in terms of the student's prior knowledge of the topic, the knowledge representations generated in performing the assigned task, and the operations performed in generating links to new information.

One task, for example, required students to interpret the results of an experiment involving photosynthesis in terms of their knowledge of the chemistry of photosynthesis. Their task involved (a) comprehending the experiment, (b) retrieving relevant information from memory, and (c) generating appropriate links between (a) and (b). Protocols from different students demonstrate differences in approaches to the problem, such as forward and backward reasoning processes. Another approach to assessing performance is to display a student's structure as an overlay on a structure that represents a consensus among experts as to what constitutes an "ideal" answer.

Subjects at different grade levels or different levels of competency have been shown by such methods to differ with regard to patterns of performance in comprehending texts of different kinds (C. H. Frederiksen, 1984), and qualitative differences between novice and expert physicians in case comprehension have been identified (Patel and Frederiksen, 1984). Several states are experimenting with analogous methods for analyzing samples of student writing in state assessment programs, even without using computers to analyze individual protocols. The procedures require human judgment and are not intrinsically dependent on the computer, but computerized assistance may make the method feasible for widespread use.

There are many other possible formats, including not only tests that require written responses but also tasks requiring hands-on operation of laboratory equipment. For example, students can be given the necessary materials and equipment and asked to design and carry out models of scientific investigations that demonstrate understanding of such scientific concepts as density, conductivity, and capillarity. Such tests are already in use on a limited scale by NAEP (Blumberg et al., 1986; National Assessment of Educational Progress, 1987), IEA in the Second International Science Study (Jacobson, 1985), the British Assessment of Performance Unit Series (1983–1985), and others (Hein, in press).

The availability of microprocessor-based computers in the classroom is growing at such a rate that it is not unreasonable to assume that in the near future every classroom, from kindergarten upward, will have access to computers. (According to Becker [1986], a national survey conducted in 1985 found that between 1983 and 1985 the number of computers in use for school instruction quadrupled from 250,000 to over 1 million.) Furthermore, while costs are decreasing, processing power, ability to produce graphics displays, and mass storage capabilities are at very high levels.

The classroom computer can play a powerful role not only in evaluating learning but also in helping students learn science processes and the higher-order thinking skills involved (Goldstein, 1980; Sleeman and Brown, 1982). While the committee's chief interest is in improving assessment of student learning, it is important to consider as well the improvement of learning, given that software can be developed to serve both purposes simultaneously. Improvement in learning is made possible because of the capability of the computer, with appropriate software, to simulate real-world scientific investigations (Clancey, 1979). Ideally, such simulations should reflect hands-on science done inside or outside of the classroom. The computer can be used to provide simulated experiments that reinforce, review, and extend the hands-on studies. Simulations also make it possible to speed up or slow down the progress of time, enlarge or shrink distances, and modify or eliminate such factors as friction and gravitation.

If such simulations are integrated into appropriate host software systems, they can be powerful tools for assessment. The host software could remember the performance of each individual student on a mass storage device, such as a floppy disk; could provide the classroom teacher with appropriate summary information on the class as a whole; and could provide the option to examine in as much detail as desired the performance of individual students. A simulation might be structured with regard to levels of achievement and could grant scoring points for good performance, just as good game software does. In this way, the simulations could give students valuable feedback as they use them, as well as storing information for the use of teachers and for the assessment of schools or school districts. Thus, the same information can be used for instructional or student evaluation purposes by the teacher, for local monitoring purposes by the principal or school superintendent, and as part of a state or national data base on student learning.

As possible instruments for national assessment, simulations would provide a solution to the problem of testing for real skills in doing science. They can be the kind of tests that *should* be taught to—which by their use will generate higher-quality science instruction. It appears entirely practical to use simulations for classroom learning and to draw on a subset of the same group of simulations for local, state, and national assessment. From the standpoint of efficient use of financial and intellectual resources, this seems desirable. Since high-quality simulations are difficult and costly to create, it is important to maximize their use once they are in place. It is also more likely that better testing methods will be developed if at the same time they can be used to improve instruction.

Assessment of Conceptual Knowledge and Processing Skills

Cognitive scientists, including both psychologists and computer scientists working in the area of artificial intelligence, are developing models of intellectual functioning that have relevance for assessment (Bransford et al., 1986). Cognitive scientists view students as information processors who possess a variety of capabilities that enable them to learn and function intelligently. These include the development of conceptual knowledge—organizing information according to structures or frameworks appropriate to the subject matter so as to give it meaning or, in mathematics and science specifically, imposing meaning on formal symbols and rules (Resnick, 1987). For example, in the sciences, the way and the extent to which scientific principles are used to organize perception, problem solving, and reasoning distinguishes the novice from the expert.

The development of conceptual knowledge is supported by specific processing skills that assist in the absorption of information and its organization and use; they include processing speed, memory capacity, memory organization, factual knowledge, and procedural knowledge (Kyllonen, 1986). Procedural knowledge includes not only knowledge of algorithms but also the ability to plan and use various heuristics and strategies. All these capacities function interactively in contributing to learning and intelligent behavior. An understanding of how they function should facilitate instruction (N. Frederiksen, 1984b), and an ability to assess these capabilities should be valuable not only to teachers and curriculum designers but also to educators at state and national levels.

This information-processing conception of learning and intellectual performance is too complex to describe here. What follows are brief descriptions of a number of possible assessment procedures aimed at certain cognitive abilities, ordered roughly according to the complexity of the ability and the difficulties involved in assessing it. The procedures suggested are generally based on experimental methods that have been devised by cognitive scientists for research purposes. Few of the procedures have been used for assessment, and much work will be needed before they can be used systematically in assessing proficiency in science and mathematics.

Speed of Processing Processing speed is typically measured in terms of response latencies (reaction time) in performing acts that are relevant to an area of expertise. For example, in learning to read, the beginner must learn how to translate letter combinations into speech sounds and to relate those sounds to words stored in memory. These may be difficult tasks for a young child, but for a skilled reader they are performed very quickly and without attention. It has been shown that differences in response latencies in word analysis, discourse analysis (e.g., identifying the antecedent of a pronoun), and integrative processes (e.g., generating extrapolations from the text) distinguish the proficient reader from a less skilled reader (J. R. Frederiksen, 1982). Speed is important as an indicator because it shows that a process can be carried out automatically, without attention, and therefore does not interfere with other more complex mental processes that are going on simultaneously (Schneider and Shiffrin, 1977; Shiffrin and Schneider, 1977). In the case of reading, "... automaticity of word-analysis skills essentially frees processing resources for the purpose of discourse analysis" (J. R. Frederiksen, 1982:172) and "... these skills are poorly represented in conventional tests of reading comprehension" (p. 173).

The need for automatic processing in elementary arithmetic is well known to teachers (although probably not by that term), and they try to increase automaticity by such means as drill with flash cards. Use of a computer would facilitate such training and would also make it possible to measure response latencies and, thus, identify those instances of finger counting or some other "short-cut" method that actually increases response time. In algebra, automatic processing could be assessed by having the student carry out simple transformations of equations and measuring the response latencies.

Moreover, patterns of latencies have been used to distinguish what kinds of procedures children use for addition and subtraction, for example, and how students and experts break algebraic equations into meaningful units. Thus, speed measures are useful not only for assessing automaticity but also for monitoring procedural skills.

Pattern Recognition Pattern recognition is a skill related to speed of processing. With much practice one can learn to recognize very quickly a complex stimulus that may be embedded in a still more complex background. This phenomenon was first observed by deGroot (1965) in comparing chess grand masters with ordinary chess players. He found that grand masters were able to reproduce correctly the positions on a board of 20 to 25 chess pieces in a midgame position after seeing them for a few seconds, while ordinary players could reproduce correctly only a half-dozen pieces. Apparently grand masters had learned after years of staring at chess boards to quickly perceive and use patterns in processing data. Simon and Chase (1973) and Simon (1974) later timed the placement of the pieces and found that the intervals between placements were relatively short for the pieces in a cluster and that longer intervals defined the boundaries between clusters. Similar pattern-recognition skills have been identified in recognizing functional elements (e.g., stages of amplification) in a schematic by electronics experts (Egan and Schwartz, 1979) and in identifying the important signs and symptoms of a disease by experienced physicians (Barrows et al., 1982). Pattern recognition is important in many activities, and measures of this skill might be an indicator of proficiency because, like automaticity, such skill reduces the load on working memory and makes its resources available for other, more complex activities. Measuring latencies in responding to relevant tasks would be an appropriate method for assessing a pattern-recognition skill.

Organization of Knowledge How knowledge is organized in long-term memory may be another useful indicator of an aspect of information processing. The elements in long-term memory are items of information and clusters of such items, which are interrelated in complex ways to form an extremely large system. The organization may involve temporal, spatial, hierarchical, causal, and other kinds of relationships. Presumably the organization depends on the number and kinds of experiences one has had with the elements, and retrieval would depend on the strength of their interrelationships (Hayes-Roth, 1977; Gentner and Gentner, 1983). Highly organized cognitive structures are formed as one acquires expertise in an area such as mechanics or forestry. Since accessibility of stored information depends on how it is organized, it would undoubtedly be useful to know how information is organized in the minds of students and how that organization changes with practice.

One cannot hope to discover how all the information in memory is organized, but methods are available for assessing the structure of knowledge in particular domains. One method is to ask students to recall items of information and to time the responses—a method analogous to that used to investigate the size and nature of clusters of chess pieces as perceived by grand masters. Sets of closely related items tend to occur with short latencies, while longer intervals tend to mark the boundaries between sets. Another method is merely to have students sort the elements into clusters.

A more sophisticated method makes use of judgments of similarity between pairs of words that represent the key concepts in a domain (e.g., in mechanics, such words as mass, force, velocity, acceleration, density, volume). A student's ratings of all the possible pairs is analyzed by multivariate scaling, which produces a multidimensional representation of a structure. This structure then can be compared with that obtained from the judgments of experts (Shavelson, 1972, 1974; Meyer and Schvaneveldt, 1976; Preece, 1976; Diekhoff, 1983; Schvaneveldt et al., 1985). The structure based on the judgments of experts in physics was found to fit a structure based on physical theory, and student structures were found to improve with instruction in physics (Shavelson, 1985). Thus, it seems feasible to develop for a variety of subject-matter areas assessment methods that provide some information about the organization of information in memory for individuals or for groups of students.

Skill in Retrieving Information The accessibility of information stored in memory has for many years been assessed by means of aptitude tests presumed to measure the fluency with which associations and ideas are generated. The ability is very general and is thought to be related to creativity. It is possible that analogous tests would be useful in certain specific domains of expertise to elicit responses related to particular topics in that domain. Students of botany, for example, might be asked such questions as "What might be the cause of the fruit dropping from an apple tree before the apples are ripe?", and the test might be scored in terms of number and quality of the ideas.

Internal Representations of Problems How students conceive of a problem has much to do with their success in solving it. A given student's representation or mental model might take the form of a set of verbal propositions, a spatial arrangement of the problem elements, a picture, a chart or diagram, an equation, or an algorithm (see Larkin, 1979; Larkin et al., 1980). If a crucial element is omitted or if the representation is inaccurate, solving the problem will be difficult or impossible. It would be useful to know what problem representations are used by students when they attempt to solve a certain type of problem.

The most commonly used method in studying problem-solving behavior is the "think aloud" method of collecting protocols, in which students are instructed to report what they are thinking as they attempt to solve a problem (Newell and Simon, 1972; Ericsson and Simon, 1984). Once a protocol is obtained, it may be interpreted in terms of the cognitive processes that are involved. This type of analysis has been used with some success in mathematics; pairs of students have been videotaped as they discuss a problem on which they are working together (Schoenfeld, 1982). Methods using protocol analysis would be useful in investigating how a problem is represented internally and how that representation changes with training and practice.

Another method of studying problem representations involves asking experts and novices to sort a set of problems into categories. The results in physics, where the method has been applied, indicate that novices tend to sort the problems on the basis of superficial characteristics of the problems, such as the use of inclined planes or pulleys, while the experts categorized the problems in terms of the physical principles that were involved (Chi et al., 1981). Asking students to sort problems is a possible way of discovering something important about the internal representations of problems that they use.

Research on the misconceptions that many students have regarding physical phenomena shows the importance of discovering student conceptions of problems (Stevens et al., 1979; McDermott, 1984). For example, it has been shown that some children believe that they are able to see an object because their vision goes from the eye to the object, rather than because light from the sun is reflected by the object to the eyes (Anderson and Smith, 1983; Anderson, 1985). And it is reported that an appreciable number of students, even those who have had a course in physics, believe that when an object is released from the rim of a spinning wheel it will follow a spiral trajectory in space. Such misconceptions have been shown to be so enduring that some students reinterpret statements of physical laws to make them consistent with the misconception. Misconceptions about physical phenomena often can be discovered by asking a student to draw or otherwise indicate what he or she thought was happening or would happen under certain conditions.

Computers have been used to assess students' understanding of physical laws. One simulation depicts a Newtonian world without friction or gravitation in which objects obey the laws of motion.

When given the task of moving objects from place to place by applying force, students are often surprised by the results, indicating inadequacies in their understanding of Newtonian physics (White, 1983). Such a simulation could be used both for assessment and for instruction.

Procedural Knowledge The term *procedural knowledge* includes not only knowledge of such routine procedures as the algorithms used in computation but also more complex skills. Complex skills may involve, for example, planning the steps to be taken in solving a problem and the use of strategies or such heuristics as means-end analysis, reformulating a problem, or thinking of analogies to a problem situation. Computer programs have been developed that make it possible to discover the erroneous algorithms ("bugs") that some students use in attempting to solve arithmetic problems (Brown and Burton, 1978; Brown and VanLehn, 1980). One well-known bug, for example, involves subtracting the smaller number from the larger regardless of which one is on top. Many other bugs have been found to exist that are unknown to most teachers. New computer programs provide detailed information about the sequence of steps (the solution path) that was taken by a student, and, from that information, the strategic errors committed because of inadequate mathematical understanding may be inferred.

Other programs are intended to discover and assess the depth of a student's understanding of an area of expertise. For example, computerized algebra tools now being developed permit students to see and manipulate the array of possible steps that they could take as they attempt to solve an algebra problem. Knowing the path students take through this "search tree" reveals much more about their skills in algebra than does the number of correct answers to the problems, including such metacognitive skills as choosing an appropriate strategy, profiting from errors, and the ability to monitor one's own performance. Similar programs are now available in other areas of mathematics, including the Geometric Supposer (Schwartz and Yerushalmy, 1985) and the Semantic Calculator (Schwartz, 1983).

Computerized coaching systems are being developed that monitor a student's problem-solving performance. Based on diagnostic models that are integral parts of the system, computer programs can be designed that offer advice to the student and at the same time provide detailed assessments of his or her capabilities (e.g., Burton and Brown, 1979; Anderson et al., 1985). Computerized medical problem-solving programs have been developed that offer to the physician not only advice but also explanations or reasons for the advice (Reggia et al., 1985). Such systems are now capable of assessing performance in very complex domains.

Another feature of the computer is that it can keep track of the collection of strategies that a student tries in solving a problem and then generate a summary of what he or she has tried and has neglected to try. Thus, the computer opens up several new possibilities for assessment. The interactive nature of the student-computer relationship allows the student's capabilities to be progressively disclosed; if the student is unable to deal with a problem, more information or hints can be given (Reiser et al., 1985). In this manner, a single problem can be used for both assessment and instructional purposes.

Not all the computerized assessment procedures described above can be administered with a microcomputer; some may require the use of a sophisticated work station. The costs of such work stations have been decreasing at a rapid pace and are likely to continue to do so. Within five years, such equipment will not be out of reach, at least for assessments on a four-year cycle. In the meantime, much can be done with small computers. As the cost of computers continues to decline, more assessments will become affordable.

A note of caution is in order. Too much reliance on computerized testing and teaching may result in a tendency to substitute computer simulation for real-world experience, or to tilt testing methodology toward those exercises that are most easily computerized. Users and creators must be alert to minimize such tendencies, and innovative assessment devices that do not require a computer should also be developed and made available.

The Development and Use of New Methods

None of the assessment methods described in this section can compete with multiple-choice tests from the standpoint of economy and efficiency, although matrix sampling makes their use more feasible. However, investment in the development of the recommended new methods and the cost of using them is, in the committee's view, justifiable not only because these methods would provide information for a far more accurate and complete assessment of instruction and student learning, but also because they are likely to be useful in the instructional process itself (see, e.g., Linn, 1986). Exercises could be used not only for assessment but also for practice and to provide information for remediation, and assessments based on exercises designed to probe higher-order learning should raise educational standards by providing models of performance to be emulated by both students and teachers.

An organization is needed to encourage, conduct, and coordinate the development of the needed assessment materials. The development of new assessment materials is costly, in both money and intellectual resources; needless duplication of effort must be avoided. This implies that the areas most in need of research and development of assessment techniques must be defined, newly developed instruments must be evaluated for their quality, and facilities for the distribution of materials to schools and teachers must be created.

The problem of test validation is particularly important for any new generation of tests that may be developed to assess proficiency in science and mathematics. The approach that has typically been used for test validation—finding a variable that may be thought of as a criterion and computing a correlation—will probably not be feasible, since no reasonable criterion is likely to exist. Clearly, another method is needed.

The most reasonable method for validating the kinds of tests that have been proposed is construct validation (Cronbach and Meehl, 1955). Messick (1975) defines construct validity as "the process of marshalling evidence in the form of theoretically relevant empirical relations to support the inference that [a test score] has a particular meaning" (p. 955). The implication is that a theory about the nature of the performance in question is necessary, and validation of a test involves a scientific investigation to see if the procedures and cognitive processes displayed in taking the test are consistent with the theory.

A study of construct validity of free-response tests intended to measure skill in problem solving may be used as an illustration (N. Frederiksen, 1986). One test consisted of diagnostic problems that simulated a meeting of a doctor and a new patient, and the other test involved nonmedical problems, such as why there are relatively fewer old people in Iron City than in the rest of the state. Both tests used a format that required examinees to go through several cycles of writing hypotheses, asking for information to test their hypotheses, and revising the list of hypotheses until they arrived at a solution. The subjects were fourth-year medical students. The theory about cognitive processes assumed that such verbal skills as reading, various reasoning abilities, science knowledge, and cognitive flexibility (ability to give up unpromising leads) would all be involved for both kinds of problems. In addition, skill in retrieving relevant information from long-term memory would be important. In the case of the medical problems, medical knowledge would of course also be necessary. The salient findings from a correlational analysis of the data showed that, as expected, medical knowledge was clearly the most important resource in solving the medical problems, and of course it was of little or no help in dealing with the nonmedical problems. For nonmedical problems, ideational fluency, or skill in retrieving relevant information from memory, was by far the best predictor of performance, but it was of little or no value in solving medical problems. Thus the information-processing theory had to be revised.

Embretson (1983) reports a more elaborate study involving latent-trait modeling for the identification of the theoretical mechanisms that underlie performance on a task and exploring the network of relationships of test scores to relevant variables. Experimental methods for testing a theory about test performance are also feasible and probably are preferable to correlational methods.

Summary

Currently available multiple-choice tests are adequate primarily for assessing student learning of the declarative knowledge of a subject. They are not adequate for assessing conceptual knowledge, most process skills, and the higher-order thinking that scientists, mathematicians, and educators consider most important. Since current efforts to improve curricula are beginning to concentrate on these skills, new tests and other assessment devices are needed to serve as national indicators of student learning in mathematics and science. The tests should include exercises that employ free-response techniques—not only pencil-and-paper problems but also hands-on science experiments and computer simulations. Tests for measuring the component skills involved in reasoning and problem solving should also be developed. The improvements in testing can be made feasible, despite higher costs, by the use of computer-based techniques, by matrix-sampling methods, and by the use in instruction of exercises developed for the tests.

Currently the area of greatest curricular change is in elementary school, grades K–5. A number of school systems are attempting to implement inquiry-based, hands-on instructional programs in science. These programs are considered exemplary by both scientists and science teachers, and they urgently need the support of assessment instruments that match the new emphasis on teaching for understanding and for more complex thinking skills. Prototypes of free-response techniques exist that could be adapted for use at the K–5 level in the near future.

Recommendations

Indicators of student learning at the national, state, and local levels should be based on scores derived from assessment methods that are consonant with a curriculum that includes all major curricular objectives, including the learning of factual and conceptual knowledge, process skills, and higher-order thinking in specific content areas. Such tests should exhibit a range of testing methodology, including use of free-response formats.

Research and Development: To provide the requisite tests for use as indicators of student learning, the committee recommends that a greatly accelerated program of research and development be undertaken aimed at the construction of free-response materials and techniques that measure skills not measured with multiple-choice tests. The committee urges that the development of science tests at the K–5 level receive immediate attention.

Techniques to be developed include problem-solving tasks, as exemplified by the College Board Advanced Placement Tests; pencil-and-paper tests of hypothesis formulation, experimental design, and other tasks requiring productive-thinking skills, as exemplified by questions in the British Assessment of Performance Unit Series; hands-on experimental exercises, as exemplified by some test materials administered by the National Assessment of Educational Progress (NAEP) and the International Association for the Evaluation of Educational Achievement (IEA), and simulations of scientific phenomena with classroom microcomputers that give students opportunities for experimental manipulations and prediction of results.

The creation of new science tests for grades K–5 should be done by teams that include personnel from the school districts that have been developing hands-on curricula to ensure that the new tests match the objectives of this type of instruction. In addition to providing valid national indicators of learning in areas of great importance, such new assessment materials for science in grades K–5 will provide models of tests that state and local school officials may want to adopt and use.

Key Indicator: The committee recommends that assessment of student learning using the best available tests and testing methods continue to be pursued in order to provide periodic indicators of the quality of science and mathematics education.

Tests should be given to students in upper-elementary, middle, and senior high school (for example, in grades 4, 8, and 12). Because of the rapid changes taking place in science instruction in grades K–5, assessment at this level should be carried out every two years, using exercises developed according to the preceding recommendation. For higher levels, a four-year cycle is appropriate. The tests should be given to a national sample, using matrix-sampling techniques. Test scores should be available for each test item or exercise and should be reported over time and by student subgroups (e.g., gender, race, ethnicity, rural/inner city/suburban community). As in previous assessments, results should also be reported by geographic region; efforts now under way may make possible state-by-state comparisons in the future. Similar procedures are appropriate for indicators of state or district assessments of student learning.

Research and Development: The committee recommends that a research and development center be established to provide for the efficient production, evaluation, and distribution of assessment materials for use as indicators of student learning at district, state, and national levels and for use by teachers in instruction.

The center should function as a centralized resource and clearinghouse that would make it possible for school people to survey the available assessment materials and obtain those desired. It might be called the National Science and Mathematics Assessment Resource

Center. It should be tied closely to efforts to improve the curriculum and be an active partner in the total system of educational reform. The committee suggests that as a beginning a group of experts be convened to prepare a plan for the creation of the proposed center, including its management and operation, and that the plan serve as the basis for the founding of the center by a suitable educational establishment or a consortium of universities and educational research organizations.

IMPLICATIONS FOR STATE EDUCATION AGENCIES

The assessment of what students have learned and their ability to apply that knowledge is a major task of accountability for state education agencies. Such assessments can function to assure the public and their elected representatives that both human and material resources are available and meet certain standards, that the resources are appropriately distributed to schools, and that the effects of all the human and monetary investments are reflected in student learning.

Using that basic premise, the state has a vital stake in valid yet feasible ways to evaluate what students know about mathematics and science. The state's role of leadership in assessment is quite important, and the committee is concerned that the complexities of assessing student learning be clearly understood and then attacked. If the state language-arts assessment is merely a multiple-choice grammar test, a direct message (intended or not) is sent to every teacher that the writing process itself is not important. Similarly, in the committee's view, if a state or school science assessment consists solely of a multiple-choice test, then clearly the measurement is equally limited.

Representatives of state and local systems told the committee that the recommended assessment resource center, if it were to be implemented, would fill a major gap for schools, states, and the Assessment Center Project of the Council of Chief State School Officers (see Appendix D). The assessment approaches based on hands-on investigation and computer simulation that would evolve from the proposed resource center could serve two functions for states and local communities. On a sample basis, the results of assessments using such new techniques would themselves be an important indicator at the state and national levels of student learning, and simultaneously such an assessment approach would provide a model that the committee believes to be important. While states may be able to contribute to the assessment resource center, probably only a nongovernmental institution could muster sufficient resources to develop and evaluate the new approaches, as well as to create imaginative ways to improve traditional multiple-choice testing of factual knowledge and simpler kinds of procedural knowledge. The curriculum frameworks discussed in Chapter 7 should guide the development at the proposed resource center of outcome measures, including measures not only of factual and conceptual knowledge, but also of the information-processing skills that are necessary for acquiring proficiency in science and mathematics.

ASSESSING ADULT SCIENTIFIC AND MATHEMATICAL LITERACY

There are several reasons why assessment of student learning should be extended to assess trends in the science and mathematics literacy of the entire population. First, one of the reasons to care about the quality of mathematics and science instruction in school is that it will influence mathematics and science literacy throughout the population; trends in the mathematics and science literacy of adults will in time provide information about the long-term consequences of attempts to improve the science and mathematics education provided in the nation's schools. The issue of adult literacy may raise important questions as to whether schools should emphasize immediate knowledge retention or learning that is likely to be retained in adulthood. Second, children's interest in mathematics and science is influenced by the extent to which the adults in their lives know about and show an interest in these subjects. Consequently, changes in the science and mathematics literacy of adults may foster changes in the skills and attitudes about science and mathematics that students bring to school. Third, and most important, the science and mathematics literacy of adults is a major goal of science and mathematics education.

Results of previous efforts to assess scientific literacy in the United States have not been reassuring. For example, Miller (1986) reports on surveys of U.S. adults conducted in 1979 and 1985 that included questions on the meaning of scientific study, cognitive science knowledge, and attitudes on organized science. On the basis of the survey responses, he classified 7 percent of the public to be scientifically literate in 1979 and 5 percent in 1985. Young adults (ages 17–34) did slightly better (11 and 7 percent, respectively); also, the percentage increased with increasing education. However, within the population who were high school graduates but who had not gone on to college, only 2 percent in 1979 and 3 percent in 1985 were deemed to be scientifically literate. Such results increase the need for future study of the population's scientific literacy and the long-term effects of science education.

Desired Attributes of Indicators

Any plan to generate indicators of scientific and mathematical literacy should try to estimate the degree to which a population possesses the kind of knowledge and intellectual skills outlined in Chapter 2. Assessment plans should be based on the following considerations:

- A single measure will not do, because science and mathematics literacy involves multiple dimensions of a complex set of characteristics. The indicators to be used should be matched to the models of literacy discussed in Chapter 2.
- The indicators should recognize that there is no single, absolute level of literacy and that various levels of attainment in different components of a community or population group are likely.
- Any measures used to generate indicators should be supplemented by research to validate what is actually being measured.
- Indicators may be expressed in terms of descriptive patterns of problem solving and other nonnumerical ways.

At this stage, there is no particular reason to favor one method of data collection over another. Therefore, several techniques, such as conducting surveys (see, e.g., Miller, 1983), interviews, and case studies, should be considered in deciding what information to collect in order to develop indicators. As with students, traditional methods may work reasonably well to assess knowledge, but indicators should also probe the population's understanding of the nature of science and its role in society. It is particularly important and difficult to obtain reliable estimates of problem-solving skills. In the committee's view, their assessment must go beyond individual pencil-and-paper tests and should include observation and analysis of individual and group responses to carefully selected phenomena involving real objects and filmed sequences of events.

In some sense, the need for assessment of adult literacy is not as urgent as the need for assessing students; after all, fewer policy decisions will or can be driven by such assessments. Therefore, the next two or three years can be devoted to the interim development of pencil-and-paper tests and tests involving real objects. These will provide a measure of adult literacy that can be correlated both to existing tests of learning (say, of 17-year-olds) and to the assessment techniques that the committee has proposed for in-school learning. Since an important aspect of science and mathematics literacy is continuing self-education, some of the assessment techniques suggested in the preceding sections may also be appropriate for adult literacy.

Target Populations for Assessment

The committee considers education policy makers for elementary and secondary schools at state, local, and also national levels to be prime users of indicators of the quality of science and mathematics education. This has implications and raises interesting issues for the design of a set of indicators to assess the scientific and mathematical literacy of adults. One issue, for example, is how the out-of-school population should be stratified in various ways for assessment purposes. One way is to divide it into the following groups:

- Parents and guardians of children enrolled in elementary/secondary schools, public or private; alternatively, those with school-age children.
- Individuals who work in mathematics- or science-related fields or use mathematics or science in their work.
- Individuals, stratified perhaps by age groups related to other national surveys, such as the National Assessment of Education Progress and the longitudinal follow-up surveys of earlier high school classes sponsored by the Center for Education Statistics (National Center for Education Statistics, 1981, 1984).

Considering the first group, if an in-school science assessment includes a particular student, should the parents or guardians of that student be included in a science literacy assessment? If so, should the assessment include both parents, a randomly selected parent, the mother, the father, or some combination of these?

Data Collection Strategies

The following suggestions outline an initial program and illustrate one way in which a measurement effort might begin. The agency assigned responsibility for the measurement of scientific and mathematical literacy should be given responsibility for developing the details of methodology. An initial program would be devoted to:

- providing benchmark data for the country as a whole, using largely available material, and
- research to develop, validate, and field-test instruments to better measure people's understanding of the nature of science and to obtain reliable estimates of their problem-solving skills.

The projected interviews and administration of exercises probably would require personal visits to households by the interviewers, although some screening of households and some data collection might be done by telephone. The assessment might begin by providing benchmark data for all adults by gender and by broad age group and for parents and guardians of school-age children. This data base would later be expanded to provide measures for subgroups of the population, for example, by educational attainment and by race and ethnicity.

Although the program would be targeted to adults 17 years of age and older, it should be expanded to include children in elementary and secondary schools as in-school testing programs begin to include the measurement of scientific and mathematical literacy. The objective would be to provide links between school and household measures, as well as to provide a household-based unit of analysis for adults and children.

If the assessment is to serve as a reliable base for policy, it will need to be based on a probability sample for which estimates of statistical reliability can be provided. The goal should be a high rate of cooperation in the survey by individuals selected in the sample. Completion rates of from 85 to 90 percent are a reasonable expectation. The sampling could be based on a multistage approach. At the first stage, a sample of perhaps 100 areas would be selected. These areas would be counties or school districts and, if spread proportionately across the country, would be distributed across about 40 states. The sample could, however, be designed to include all states. Within each area, a sample of no fewer than 50 adults would be drawn from randomly selected city blocks or corresponding small areas outside cities, with at least 5 households sampled per block and 1 adult interviewed per household. Households would be sampled for this purpose according to their number of adults in order to give each adult tested approximately the same weight. With an 80 percent cooperation rate, this plan would yield interviews/tests with no fewer than 4,000 adults. That number would provide an adequate data base for analysis.

To monitor changes in the population, the basic survey should be repeated at four-year intervals. During intervening years, effort could be concentrated on developing and testing improved assessment methodology.

Assessing Grasp of Grand Conceptual Schemes

As with school students, it is important to find out to what extent the adult population is familiar with key scientific concepts and understands their applications. While such high-order knowledge may seem at first to resist assessment, it can be probed with the following kind of exchange, probably best administered in an interview:

- Listen to (or look at) this list of ideas: plate tectonics, evolution, gravitation, the periodic table. Is there one of them that you would be willing to talk about a bit more?
- Response (for example): plate tectonics.
- I'd like you to take a few minutes to think about plate tectonics. Please think about these two questions and answer them in whichever order you prefer. How would you briefly describe what plate tectonics are to someone who didn't know about them? What examples can you give me of things or events that plate tectonics cause or are involved in?
- Would you like to talk about another of these ideas?

Several aspects of this sample exchange are important. First, it is in the free-response format, which is needed to probe the active knowledge of the respondent and to permit flexibility in answering. Second, it evokes both a definition and specific applications of the selected "grand conceptual scheme." Since part of the power of these schemes is their ability to unify phenomena, being able to define the terms without appreciating any of the applications is to lose much of their force. Third, by including some example that virtually all adults have encountered, a minimum level of literacy can be assessed. Finally, because the questions are open-ended and recursive, they permit assessment of both breadth and depth. Although it may be difficult to do so, it would be important to establish to what extent people's responses are based on knowledge gained in school and to what extent they draw on knowledge gained from subsequent reading, television programs, museum visits, and so on, even given that school ought to teach one to continue to learn beyond one's formal education.

Recommendation

Key Indicator: The committee recommends that, starting in 1989, the scientific and mathematical literacy of a random sample of adults (including 17-year-olds) be assessed. The assessment should tap the several dimensions of literacy discussed in Chapter 2 and should be carried out every four years.

To make the desired types of assessment possible, effort should be devoted over the next two years to developing interim assessment tools that use some free-response and some problem-solving components; these assessment tools should be used until more innovative assessment techniques, described in this chapter, are available. The data collected should be aggregated to provide measures of depth and breadth of knowledge and understanding. They should also be aggregated by age, gender, race, ethnicity, socioeconomic status, and geographic region so as to establish to what extent there are systematic inequities in the distribution of scientific and mathematical literacy.

5

Indicators of Student Behavior

STUDENTS AS KEY ACTORS

At the very center of science and mathematics education lies the behavior of students. It is these behaviors that allow society to gauge and compare the extent to which the educational system is providing opportunities to learn and nurturing the attitudes important for a scientifically literate society. Moreover, a focus on what students do, assuming they have choices, offers a great advantage: student behaviors may be viewed as manifestations of a large number of hard-to-measure influences on the learning of and interest in science and mathematics. It is easier to design indicators that capture the combined effect of poorly understood influences than to assess those factors separately and directly. For example, it is relatively straightforward to collect information on the number of students who choose to study physics. The decision to study physics is a clear behavioral event, an observable activity that sums up the effects of parental suggestions, guidance counselors' advice, college admission requirements, the reputation of the local physics teacher, and the student's own interest in science. Similarly, it is possible to collect data on students' reactions to science as a career without fully understanding the variety of factors that are likely to affect those reactions. These influences are important to understand, and more research is needed to understand better their role in shaping behavior. Nevertheless, it is possible to design indicators of the important behaviors of students without completely understanding the influences that cause the behaviors or the personality characteristics that the behaviors represent.

Recent research in cognitive science (Resnick, 1983) and the growing acceptance of generative or constructionist psychology (e.g., Osborne and Wittrock, 1983; Watts and Gilbert, 1983) further highlight the importance of the student in the learning process. The current view of the student learner is one who actively constructs his or her own meaning, rather than serving as a passive receptacle of the teacher's transmitted information. The constructionist view of the learner places great importance on the prior knowledge of the student and the nature of the learning activities in which the student engages. Because learners have some control over the nature and quality of their efforts, some of the responsibility for learning outcomes shifts from the teacher to the student. This gives added importance to the monitoring of student behaviors.

Welch (1984) argues that the methods of effective scientific investigation provide a model for effective science learning. The methods for learning science should follow the methods for doing science. The model—consonant with our earlier definition of scientific literacy—suggests that successful students must participate in certain activities, such as observation and experimentation; be guided by a number of beliefs about the process, such as objectivity and tentativeness; and possess certain personal traits, such as curiosity and commitment. This is not to argue that all science learning will easily and naturally flow from the hands-on activities that can be carried out successfully by students, given that they are also expected to learn complex scientific principles not easily derived from experiments that are possible in the school laboratory. It does imply that students learn much of the core of science and mathematics more effectively by emulating the behavior and habits of mind of scientists and mathematicians. If one accepts this assumption, the development and monitoring of indicators of these behaviors clearly is pertinent to assessing the health of science and mathematics education.

The previous report of this committee (Raizen and Jones, 1985) adopted an input-output model to help categorize the various elements of the domain of science education. In this model, student characteristics, such as motivation, are viewed as antecedent or input conditions. Student interactions with teachers, peers, and curriculum materials are viewed as transactions, while changes in student attitudes and achievement are seen as the outputs of the system. For example, a student with a strong interest in mathematics (antecedent) who does homework in algebra (transaction) is likely to develop an understanding of algebra (outcome). By contrast, a student with little prior knowledge of proportions who daydreams during half the chemistry class will be unlikely to become interested in a career as a chemist or learn much about the synthetic materials that make up much of the environment.

Unfortunately, this model is limited in that cause-effect relationships are difficult to define. Do students daydream because they do not understand, or do they not understand because they daydream? Should strong interest be considered an antecedent to the process or an outcome of a prior learning situation? There is a circularity among antecedent, process, and outcome that is difficult to resolve. This is one of the reasons that the committee, in the course of debating these matters, decided in this report to focus on the important behaviors of students and teachers as the major actors involved in the process of learning and not be too concerned with the categorization of these behaviors in an input-output model. Taking an algebra course, a student activity, is a desired behavior whether it is viewed as evidence of an interest in mathematics or is seen as a precursor to achievement. Paying attention in class, once a course is selected, is another level of behavior that bears on achievement.

In order to highlight the importance of student behaviors in science and mathematics and provide some structure for recommending indicators, this chapter differentiates among three categories of behaviors: (1) student activities, (2) attitudes toward science and mathematics, and (3) scientific and mathematical habits of mind. Student activities are the observable actions of students, or adults for that matter, that have been demonstrated to be important in attaining some modicum of scientific and mathematical literacy, whether or not one wishes to infer some underlying affective trait. An example is homework. Doing homework, when effectively administered and carried out, is important in student learning (Husén, 1967; Walberg et al., 1986) quite apart from its relationship to an attitude on the part of the student. So too is taking trigonometry courses or studying science an hour per week in fifth grade, particularly if the instruction is of high quality.

Attitudes toward science and mathematics are emotional reactions to the various components of these enterprises. They are personal response tendencies, developed through experience, that can be characterized as favorable or unfavorable. However, these attitudes can be inferred only from the behaviors of people. Responding "yes" to the statement, "I would like to become a mathematician when I grow up," leads one to believe that that person has a positive attitude toward mathematics. The elements of the field as perceived by the individual become the stimuli that prompt the behavior. It is the behaviors rather than the attitudes that are observed and measured and that can become indicators of the state of science and mathematics education.

The third category of behaviors derives from scientific attitudes or scientific habits of mind, as discussed in Chapter 2 in defining scientific and mathematical literacy. The behaviors involve a set of beliefs and assumptions about the natural world, certain ways of thinking, and techniques for confronting and solving problems. They are a code of ethics observed by the scientific community that has developed as part of the success pattern of science and that provides boundaries for the actions of scientists. The code includes certain characteristics of the process of doing science—objectivity, skepticism, replication of results, parsimony and elegance of concepts, theories, and proofs. It also mirrors the characteristics of successful scientists and mathematicians, for example, curiosity and commitment.

Again, one cannot measure these traits directly. One observes behaviors that are believed to be manifestations of the traits. As an example, a student who tests the accuracy of the weather predictions in the *Farmer's Almanac* by actually observing and recording temperature and precipitation each day over a period of time is believed to possess the scientific attitudes of skepticism and belief in the value of evidential tests. The traits are inferred from the behavior.

Indicators reflecting how students behave and what they believe need to be gathered concurrently and integrated with the other indicators described by the committee. This is of particular importance to individuals at the state and local levels in a position to influence what happens in schools. Unless information about behavior and attitudes is known in addition to test scores for a reasonable sample of the student population, the focus for such policy makers remains blurry and decisions tentative. For example, even if students are performing well in their curricular areas in elementary school, early warning signs of negative attitudes or behavior could predict lessened interest in high school and college mathematics and science.

STUDENT ACTIVITIES

What students do is likely to have an impact on what they learn. Because of this relationship, it is important to develop and monitor regularly indicators of selected student activities, both those conducted within the science and mathematics classroom and those likely to occur outside the school. Concurrently, however, there is a need to continue to examine the relationships between the indicators suggested below and measures of student learning. Some research using national assessment data in science indicates linkages between student learning and indicators of such behaviors as doing homework, course taking, and out-of-school science experiences (hobbies and clubs, science projects, museum attendance, extracurricular reading) (Hueftle et al., 1983; Walberg et al., 1986). However, this work needs to be updated and replicated using more recent information, for example, the new NAEP data currently being analyzed by the Educational Testing Service. Additional factors related to learning may also be discovered that may become important indicators in the future.

In-School Activities

The relationship between instructional time and student learning was discussed in the committee's earlier report, leading to recommendations on monitoring course enrollment for both science and mathematics in secondary school and instructional time in elementary and middle school (Raizen and Jones, 1985, Chapter 4). The committee still considers these measures of student behavior—whether courses and instruction are imposed on the students through school requirements or elected by them voluntarily—to be important indicators of educational quality because of their well-established effects on student achievement. We also reiterate the caution raised in the earlier report that course enrollment data, to be meaningful, must include some sort of typology or descriptive information that allows classifying the courses as to level of subject matter covered.

Course enrollment data should be obtained in enough detail to make it possible to describe the total number of pupils enrolled in specific mathematics and science classes, as well as to describe the amount of science taken by a typical student. This requires not only obtaining school-level data on science enrollments but also monitoring individual course-taking patterns. The data may be aggregated in different ways to answer such questions as: What is the average number of mathematics courses taken by graduating boys and girls? What percentage of 12th-grade students are presently enrolled in a first-year physics course, and how many of these are minority students?

At the elementary- and middle-school level, information on instructional time plays a somewhat different role in mathematics than in science. Since mathematics has an established place in the curriculum, the question of interest concerns variations in time among classrooms and schools (see, e.g., Berliner, 1978), whereas for science, particularly in grades K–6, the more important question is whether science is taught at all. Data (e.g., minutes of instruction per week) may be gathered for individuals or at the school level. The former provides such statements as: "The average third grader received 34 minutes of instruction in science each week," while the latter yields such information as: "The average school allocated 41 minutes per week to instruction in science." The latter number is likely to be larger because of student absenteeism or being out of the class (e.g., at the library or in a special reading group) when the science instruction is offered. This is particularly a problem for specific groups. For example, children from low-income homes may receive less actual instruction as a result of missing school. Another problem to which time surveys must be sensitive is the possible double-counting of homework time—as both homework and instructional time—when homework is done during school time rather than at home.

Such time-based measures of exposure to subject matter, though informative, are a mere beginning, however. We have argued above that, in order to learn science or mathematics, one must be engaged in the process of actually doing science or mathematics. From that proposition, it follows that the quality of students' classroom experiences is as important if not more so than the amount of time spent on a subject (Brophy, 1986; Brophy and Good, 1986; Good and Weinstein, 1986; Stevenson et al., 1986). Therefore, information should also be gathered on students' use of class time, that is, on what they actually do during the time periods reported as instruction. In mathematics, systematic procedures have been developed for recording observations and identifying the cognitive levels of classroom instruction and behavior (Burkhardt, 1986). In science, we have argued, quality entails the modeling of the behaviors of successful scientists. For example, how much time is devoted to hands-on scientific activities, how often do students exhibit curiosity about their world or are given opportunity to do so, and how meaningful to them are the problems that they are asked to solve? The construct to be investigated is the extent to which students are participating in the processes of science.

Studies of how students use class time should be conducted by trained observers. The observers should assess the extent to which students are engaging in laboratory activities or similar hands-on experiences that entail making observations and taking measurements of natural phenomena, doing experiments or exercises that pose problems that capture students' imagination, working alone and in teams seeking answers to questions they themselves have formulated about the world around them, communicating the results of their investigations by the written and spoken word, and questioning their findings and seeking verification by gathering additional evidence.

Information on course enrollment and instructional time and on student use of time should be collected in regular four-year intervals. The suggestion for a four-year cycle is based on several considerations: a study usually requires two years for data to be gathered, analyzed, and reported; one study should be completed before the next one on the same subject is planned; and studies conducted every four years will be frequent enough to be useful to policy makers while keeping costs and response burden within bounds. The four-year cycle also matches that proposed for testing in Chapter 4, making it possible to investigate the relationships between student behaviors and achievement. Gathering data simultaneously on teacher behaviors, student behaviors, curriculum, and achievement (see Appendix E) would provide a rich source of information for conducting research on how to improve science and mathematics education. For example, the influence of student, teacher, and curriculum on student learning could be examined in addition to exploring in some depth the variation in the effects for various subgroups based on ethnicity, race, gender, and type of community.

Out-of-School Activities

Recent research (Fraser et al., 1986) suggests that out-of-school activities are more highly correlated with science learning than are in-school activities. Hence, it is important to consider such activities when monitoring the status of science education. In mathematics, with its hierarchical structure, out-of-school activities—participation in mathematics contests, computer work, jobs that require mathematics skills—appear to be relatively less important. For example, differential course taking among high school students accounts for more than a third to over half of the variance in mathematics achievement as currently measured (Welch et al., 1982; Jones et al., 1986).

An important out-of-school behavior is the amount of homework time spent on science and mathematics. There is an accumulation of research evidence that supports the value of homework in learning a subject, particularly if homework is checked and discussed (for a summary, see Raizen and Jones, 1985:89-91). For example, Fraser et al. (1986) found that, for science test scores of 13-year-olds, and with other factors held constant, an increase of one hour per day in the time spent on homework is associated with a 7 percent increase in number of test questions answered correctly; the gain increases to 10 percent for 17-year-olds (Walberg et al., 1986). However, most research on homework deals with general amount of homework done; data on the amount of homework devoted to such specific subjects as science have seldom been systematically gathered. Data need to be gathered and analyzed not only for specific subjects but with course-taking held constant, since the amount of homework done will probably vary with the number of courses taken.

Other out-of-school behaviors that have been hypothesized to affect student learning include exposure to or involvement in (1) informal science learning situations at zoos, museums, science fairs, and the like; (2) time spent applying the content and processes of science and mathematics to one's daily life, for example, deciding on over-the-counter medication, taking certain health measures or risks like exercising or smoking, judging the veracity of a television commercial, or checking a restaurant or grocery bill; and (3) active participation in using knowledge of science and mathematics to address recurring societal problems, even in a limited way, for example, turning off lights when leaving a room or limiting the length of a shower during water shortages (conservation), maintaining a reasonable speed limit (safety and conservation of energy resources), picking up litter or turning down the volume on the radio (combatting pollution), or emptying the ashtray of the car on the street when stopped at an intersection (pollution).

Recommendations

The following three types of measures should be used as key indicators of students' in-school behavior. Data on each of the measures should be gathered and reported for gender, ethnicity, race, type of community (urban, rural, suburban), and grade level as well as by district, state, region of the country, and nationally. If discrepancies among groups continue to be found, as they have in the past, they will have important policy implications for achieving scientific and mathematical literacy for all students. The three types of measures have to do with course enrollment, time devoted to science and mathematics, and quality of instruction.

> **Key Indicator:** The committee recommends that data on secondary school course enrollment be gathered on a four-year cycle for both mathematics and science. The specific data to be gathered are the number of semesters of science and mathematics taken by students and total enrollment in the variety of science and mathematics courses offered in secondary schools.

Courses should be identified as to level of difficulty (e.g., for eighth-grade mathematics: remedial, typical, enriched, algebra). The indicators to be constructed from these data are the average number of mathematics and science courses taken and the percentage of students enrolled in specific courses.

> **Key Indicator:** The committee recommends that the data to be gathered at the elementary- and middle-school level, equivalent to course enrollment data, be the number of minutes per week devoted to the study of science and mathematics. The indicator should also be expressed both as a ratio of all instructional time and of total time spent in school.

At each policy level—national, state, and local—experts may wish to define the minimum amount of class time necessary in each grade, particularly for science. However, care needs to be taken not to countervene, through efforts to mandate or log instructional time, the potential benefits of integrating mathematics and science instruction to some extent.

Because of the importance of possible differences among various groups—ethnic and racial, gender, socioeconomic status, and so on—we recommend that the data be collected both at the level of the school and the individual student.

Key Indicator: The committee recommends development of a time-use study involving external observers to obtain some indication of the quality of the science and mathematics instruction being received. In science classes, this would include, in addition to the teaching of conceptual and factual knowledge, the percentage of time spent by students involved in the processes of science (observing, measuring, conducting experiments, asking questions, etc.). A similar study is recommended for mathematics classes; a panel of mathematics educators should determine the nature of student behaviors sought.

Supplementary Indicator: The committee recommends the collection of information on minutes per week spent on science and mathematics homework.

The frequency and detail necessary for gathering data on homework are the same as for in-school activities—that is, the information should be gathered every four years and allow analysis by ethnicity, race, gender, grade level, and size and type of community. National data are important for comparisons over time and with other countries; states and local districts may also wish to have this information. Care must be taken that homework done in school is not double counted as both homework time and instructional time.

Research and Development: The committee recommends further research and development on possible supplementary indicators in the following three areas of out-of-school student behaviors, with the goal of clarifying their relationships to student mathematics and science learning:

- Amount of time (minutes) devoted to out-of-school science and mathematics activities, for example, going to

zoos and science museums, watching science programs on television, reading science books, playing with a computer at home, voluntarily doing science projects or mathematics puzzles.
- Percentage of students reporting that they use (apply) the concepts of science and mathematics from time to time in their own lives. One way to implement this indicator is to conduct a survey on the number of times students faced a personal decision and relied on something that they learned in science or mathematics to help them make that decision.
- Percentage of students reporting that they use the concepts of science and mathematics to help them address some persistent societal problem.

At the same time that the collection of information proceeds on the recommended indicators of in-school and out-of-school student behaviors, research should be pursued in three related areas. First, better understanding is needed of the linkages between student learning and such student behaviors as course-taking, doing homework, and participating in extracurricular science or mathematics activities. The research should be designed not only to validate current findings on the linkages of these factors to learning but also to allow for the discovery of other student behaviors that strongly affect learning. Second, more work needs to be done to elaborate the constructs of student activities and how they might be measured in order to improve related indicators. Third, factors that influence student activities need to be examined, for example, who convinces children to avoid elective courses in science or what influences the amount of homework in science and mathematics that is done. If one assumes that the behavior of students inside and outside school affects learning, then it is important to understand what determines these behaviors.

Research and Development: The committee recommends continued research on linkages between student learning and various student activities, on more effective ways of assessing activities that affect learning, and on the factors that influence individuals to engage in these activities.

ATTITUDES TOWARD SCIENCE AND MATHEMATICS

Science and mathematics educators generally espouse as a goal that students acquire positive attitudes toward the various components of the scientific enterprise. These attitudes are seen to be important as outcomes of the schooling process and for their influence on the activities in which students choose to participate—as students and in later life. Liking science or mathematics is an attitude to be learned in a science or mathematics class as an end in itself, as well as to facilitate further learning in science or mathematics and eventual career choices.

Although we argue above that spontaneous behaviors are in general more trustworthy indicators than indicators of attitudes and feelings constructed from answers to questions posed by adults, attitude questionnaires are not without some value. A multitude of attitude measures have been developed. For example, Gardner references more than 200 studies in a review he wrote in 1975. An ERIC search of science testing articles written between 1975 and 1985 (Welch, 1985) revealed that more than one-third of them were devoted to the measurement of attitudes. The last three NAEP assessments of science have included items on attitudes, and they will continue to be included in future national assessments. Approximately one-eighth of the 1986 science assessment was devoted to attitude items.

Past attempts to obtain measures of attitudes toward science have focused on such topics as like or dislike of science classes, science teachers and scientists, and positive or negative judgments about the value of science, careers in science, and support of scientific research. Results are sometimes hard to interpret; for example, 49 percent of 17-year-olds in 1982 agreed or strongly agreed that their teacher makes science exciting, and 62 percent thought that their teacher was enthusiastic, yet less than 50 percent of this age group reacted positively to questions about their science classes. In general, the percentages of positive attitudes expressed by the nation's youth toward various components of science are disappointingly low (Hueftle et al., 1983).

In mathematics, the areas investigated include relationships between attitude and achievement, the influence of parents and teachers on student attitudes, and other factors related to attitudes and attitude change (Kulm, 1980). More specifically, a sizable number of studies have investigated the effects of various attitudes on women's participation in mathematics courses and careers (Chipman et al.,

1985). Generally past studies have not succeeded in establishing a strong connection between positive student attitudes regarding the subjects themselves, teachers, classes, careers, and the like and student achievement. Three reviews of research on attitudes and performance in mathematics all conclude that there is a positive correlation, although it is small (Aiken, 1970; Crosswhite, 1972; Kulm, 1980; Bell et al., 1983). Similar results have been found for science and other subjects (Welch, 1983; Willson, 1983; Horn and Walberg, 1984). These results may stem from difficulties in interpreting the meaning of attitude measures (Gardner, 1975). Items used to assess attitudes have given inconsistent and ambiguous results (Munby, 1983), raising questions as to what is really being measured.

In part because of the ambiguous findings to date, the committee suggests further work on national indicators of student attitudes toward science and mathematics. In the committee's view, it is time to examine carefully the purpose of the attitude assessments included in the NAEP, the IEA, and other major studies, to define the domain more precisely, and to develop better measures of the attitudes that are in themselves considered important outcomes of mathematics and science education or that have been demonstrated to have strong positive effects on student learning.

Recommendation

Research and Development: Given the importance attached by science and mathematics educators to the development of attitudes that will foster continuing engagement with science and mathematics, the committee recommends that research be conducted to establish which attitudes affect future student and adult behavior in this regard and to develop unambiguous measures for those that matter most.

SCIENTIFIC AND MATHEMATICAL HABITS OF MIND

In addition to developing in students cognitive competence in science and mathematics and favorable attitudes toward these fields, their education should also equip them with scientific and mathematical habits of mind, as defined in Chapter 2. These habits of mind evince themselves in behaviors that represent certain ways of thinking about the world. The behaviors themselves are thought to be manifestations of internalized personal traits that embody the scientific and mathematical world view. An example is fate control, discussed further below, a pattern of beliefs about one's relationship to events and of events to each other.

Because scientific and mathematical habits of mind are an integral part of scientific and mathematical literacy, indicators for them should be developed, monitored, and the findings reported to educators and policy makers. We provide a brief overview of several constructs thought to be relevant and recommend that further research and development be undertaken in the area of scientific and mathematical habits of mind.

Relevant Constructs

Scientific habits of mind foster an extended milieu of beliefs about the world and one's place in it. The methods of science and the values attached to it have the power to shape an individual's sense of purpose and control over his or her own life. This sense is generally referred to as fate control. If one's sense of fate control is low, one may act as if one believed events to have few connections between them and that each event springs uninvited into one's life. The world—including one's own personal world—is unpredictable, like a game of chance or a collage of happenings over which one has little control. As a rule, a person with these beliefs displays a rather primitive level of knowledge and understanding about science. If one's sense of fate control is high, one may act as if one believed that events have roots that evolve by processes one can discover and thereby possibly influence. People with this orientation are more keenly attuned to cause-and-effect relationships and to the structure of the relationship between events and ideas.

The notion of fate control is given weight by a considerable body of research that connects children's attributions of their successes or failures to their persistence and performance in school (Seligman, 1975; Lefcourt et al., 1979; Stipek and Weisz, 1981).

The measurement of fate control or attribution of success is not nearly as well developed as the constructs themselves. Answers to items probing these constructs depend on the formats used (Stipek and Weisz, 1981); there are cultural differences that may affect not only responses but also relationships between fate control, attribution of success, and student learning; attributions that are other-directed (i.e., may be interpreted to indicate low fate control) may in fact be quite realistic (e.g., "My teacher isn't very good"). At this stage, it would seem more promising, difficult as it is, to observe and assess overt behaviors that embody scientific habits of mind than to assess these through related attitudes about fate control and self-efficacy (Rowe, 1979; Blumberg et al., 1986; Educational Testing Service, 1987). Research to clarify the unequivocal core of fate control that links to the development of a scientific world view should proceed.

In mathematics, McLeod (1986) has found low positive correlations between fate control (or such related factors as locus of control, reflective/impulsive behavior, and field dependence/independence) and student achievement. Freudenthal (1983) has defined mathematical habits of mind as including the following attributes: ability to understand and use mathematical language, ability to visualize the data and the unknowns in a problem from different perspectives, grasping the degree of precision needed for a problem, knowing when and how to apply mathematics in a given context, and being aware of one's own mathematical activities. Much work on improved measures will have to be done before it will be possible to assess the extent to which students are developing these attributes.

One of the purposes of science and mathematics education is to enable and interest students in attending to these endeavors in some form throughout their lives. However, the motivation for individuals to do so, inside and outside school, is in need of much research. If mathematics and science education succeed, then individuals will leave school understanding how to apply the knowledge and processes of science and mathematics to the questions and problems they face personally and as members of society. In that connection, four constructs are advanced as relevant to consider in developing the desired attitudes, motivation, and curiosities in all students: engagement, expectations/autonomy, connectedness, and competence.

While each of the constructs taken separately has a good deal of research underlying it, how they might act together to motivate attention to science and mathematics is not well understood. (For overviews, see Weiner, 1979; Malone, 1981; Connell and Ryan, 1984; Connell, 1985.) Which of these factors, linked in what patterns, make a difference in perceptions, motivations, and quality of involvement with scientific and mathematical ideas? And how might one go about obtaining indicators of the four constructs? Both of these questions will require considerable investigation before parsimonious indicators can be recommended that might be used routinely in the assessment of the condition of mathematics and science education.

Engagement Engagement means the active, interested involvement in learning science and mathematics and making appropriate application to real problems or situations. The opposite of engagement is disaffection, which may be manifested by inattentiveness, avoidance, rebellion, or by resort to rote learning when one does not understand. The concept of discretion is relevant here. Tasks at work or school typically have two parts: prescribed and discretionary—aspects in which the person has some latitude to make choices (Jaques, 1956). As the discretionary component increases, engagement seems to increase (e.g., Cavana and Leonard, 1985).

Expectations/Autonomy This construct encompasses the sense that one's own purposes, interests, and curiosity are being served by engaging in a particular set of activities and that, to some extent and on some occasions, one can choose from among options. The ratio of intrinsic to extrinsic motivation is high, as are performance, persistence in the face of difficulties, and sustained attention to science. For some people, science is intrinsically interesting; for others, it is not so interesting but is recognized as instrumental to other goals about which they care.

Connectedness The theme of connectedness appears central. It is the degree to which students perceive that what they are doing and how they are doing it is connected to their everyday life—in career development, in health management, in their relationship to the community, and in their roles as citizens. They also need to see that ideas within the subject hang together in some fashion that makes sense rather than as a dictionary of facts; that is, there should be some thematic character to their learning. Science is always in a state of development and change, but for the most part there is coherence within the changes. The response of students to such changes could be expected to be different depending on whether they had a thematic or a discrete (dictionary-like) organization of knowledge.

Competence Competence depends on having an accurate idea of what it takes to be a successful science or mathematics student. Success is a function in part of whether one knows what strategies are necessary to be successful and whether one possesses the strategies and the will to exercise them.

Recommendation

Research and Development: The committee recommends research to identify and validate constructs related to the continuing involvement of students and adults with science and mathematics throughout their lives. In addition to the refinement of these constructs, strategies should be explored for obtaining indicators of the relevant constructs and associated behaviors.

6

Indicators of Teaching Quality

TEACHERS AS KEY ACTORS

In its earlier report (Raizen and Jones, 1985), the committee discussed potential indicators relating both to the quantity and quality of teachers responsible for science and mathematics instruction. One of the major conclusions in that report is that "the construction of ... an indicator on teacher demand and supply is at present not feasible at the national level because of the lack of a meaningful common measure of qualification" (p. 71). At the state and local levels, standards on teacher quality vary among school districts within a state and among schools within a district—appropriately so, if the schools or districts serve student populations with different needs (Wise et al., 1987). Yet a panel, set up under the committee's aegis to develop better models for estimating teacher demand and supply, is stressing "that satisfactory models of supply and demand for science and mathematics teachers must be specific regarding teacher qualifications" (Panel on Statistics on Supply and Demand for Precollege Science and Mathematics Teachers, 1987:58). Obviously, questions concerning the adequacy of instruction in science and mathematics cannot be answered until some measures of teaching effectiveness are developed and found acceptable.

What constitutes effective teaching of mathematics or science? On what should indicators of teaching effectiveness focus—characteristics of the teachers themselves? Measures of what teachers do in the classroom? In attempting to resolve this issue, the committee devoted considerable attention to the research literatures on the characteristics of effective teachers and on the determinants of effective teaching. We found strong research support for parents' conviction that teachers matter. This support comes from studies showing clearly that children enrolled in different schools, and even in different classrooms within the same school, learn different amounts during the school year (Hanushek, 1972; Murnane, 1975; Armor et al., 1976). While these studies by themselves do not demonstrate that differences among teachers alone account for why more learning takes place in some classrooms than in others, it is reasonable to infer from these and other studies that differences among teachers are one important factor contributing to these differences in student learning.

The evidence that teachers matter led us to turn to the studies that have attempted the more difficult research task of exploring the specific characteristics of teachers and the specific teacher behaviors that are related to high student achievement. Unfortunately, we concluded that such studies (whether in traditions known as input-output studies or process-product studies) do not provide significant guidance for the development of indicators of effective mathematics and science teaching. In part this may be the case because the studies are largely based on current conceptions of teaching that emphasize the learning of procedural skills rather than the larger vision of the teacher's role set out by, for example, the Holmes Group Consortium (1984) and the Carnegie Forum on Education and the Economy (1986). It is conceivable that the research results would be different if student scores on tests of higher-order thinking skills were used to measure teaching effectiveness. This hypothesis has not been tested, however, since all existing studies have measured teacher effectiveness by student scores on multiple-choice tests that, as Chapter 4 on learning assessment explains, do not measure the full range of higher-order thinking skills. The following section explains why the results of input-output studies and process-product studies do not provide guidance for indicator development.

Findings from the Literature

One type of study, called educational production functions or input-output studies, has explored the extent to which gains in student achievement can be explained by information on teachers' demographic characteristics, education, test scores, and teaching experience. There are a few relatively consistent findings. For example, teachers with at least three to five years of experience are more effective on average than beginning teachers (Hanushek, 1972; Murnane, 1975; Murnane and Phillips, 1981), and this appears to hold true for science teachers (Druva and Anderson, 1983; Penick and Yager, 1983). A somewhat less solid finding is that teachers with high scores on tests of verbal ability may be more effective than teachers with lower scores (Coleman et al., 1966; Hanushek, 1972), although there are exceptions to this pattern (Summers and Wolfe, 1977).

While it is common to focus attention on positive findings, the dominant conclusion from input-output research is that the vast majority of the variables used to depict teachers, including sex, race, possession of a master's degree, and whether the teacher was an education major as an undergraduate, are not consistently related to teaching effectiveness, whether measured by student gains on standardized achievement tests or by evaluative judgment (see, e.g., Schalock, 1979).

A second type of research, sometimes referred to as process-product studies or studies of teaching effectiveness, has examined whether specific actions of teachers are systematically related to teaching effectiveness. In recent years, this research has provided support for the sensible proposition that students' achievement in a specific subject is positively related to the amount of in-class time devoted to instruction in the subject, as noted in the preceding chapter and the committee's earlier report. The research also supports the proposition that just as important as the amount of time allocated to mathematics, science, or other subjects is how the time is used (Weber, 1978; Evertson et al., 1980; Good, 1983). This has led to studies of how best to use instructional time, including how to develop lessons and how to manage the classroom. These studies have produced insights that are helpful in teacher education, for example, by demonstrating the importance of presenting all students with challenging work and expecting them to complete it, and making smooth transitions from one activity to another (Good and Grouws, 1979; Brophy and Good, 1986).

If the process-product research had found that teachers who develop lessons effectively and manage their classrooms well do so by engaging in particular well-defined actions, then these actions could provide the basis for indicators of teaching effectiveness. Observational techniques could be used to record the extent to which teachers of mathematics and science employ these superior techniques. In fact, such a mapping of concepts that characterize effective teaching to well-defined teaching actions has not been possible, however. Consequently, the process-product literature provides little guidance for the development of indicators of teaching effectiveness. There are at least two reasons for this: first, effective teaching requires carrying out more than one action. Carrying out requisite actions in isolation may not result in effective teaching; and, as a result, observations of the frequency with which teachers carry out a single particular action would not provide the basis for a reliable indicator. Second, the set of particular actions that results in effective teaching may depend on the type of classroom situation the teacher is in. The actions that are most effective would be expected to vary with grade level and with the subject matter and skills being taught. In addition, there is some evidence that effective teaching of children with different characteristics and backgrounds requires different sets of actions by the teacher (Cronbach and Snow, 1977; Brophy and Good, 1986). These complexities in mapping concepts to actions would make it very difficult to base reliable indicators of teaching effectiveness on observations of whether teachers carry out specific, well-defined actions (Brophy, 1986).

In summary, review of the research on the determinants of teaching effectiveness led us to the conclusion that neither input-output studies nor process-product studies provide sure guidance for the development of indicators of the quality of mathematics and science instruction in school. In one sense this is discouraging, because it makes the task of developing reliable indicators of teaching effectiveness more difficult. In a different sense, however, the results are encouraging, because they underline the fact that effective teachers cannot be defined merely as individuals with specific demographic characteristics who have earned particular academic degrees, or as people who have been trained to behave in predictable, routinized ways in the classroom. Such definitions obscure the characteristics that effective teachers have in common—the skills and attitudes of professionals (Holmes Group Consortium, 1984; Carnegie Forum on

Education and the Economy, 1986; Darling-Hammond and Hudson, 1986).

The Professional Teacher

In recent years, at least 44 states, several major commissions, and the national teachers' unions have moved toward a definition of the professional teacher. The following attributes are generally included in the definition: professional teachers understand the subject matter they teach and its relation to other subjects in the curriculum. They possess a high degree of intellectual curiosity, which is reflected in how they spend their time. Professional teachers also have the desire to help students increase their skills and self-confidence, and they have the skills to achieve these goals, including being able to adapt the curriculum to fit the needs of their students (Good and Weinstein, 1986). Finally, professional teachers continue to learn new things as they progress through their careers. It is teachers with these attributes that are wanted and needed to provide instruction in mathematics and science.

Schools cannot attract and retain professional teachers unless they provide the support that professionals need and can find in other occupations (Darling-Hammond, 1984; Rosenholtz, 1985). This support includes competitive salaries, opportunities for professional development, and significant control over the time, space, materials, and curriculum needed to teach effectively (Lightfoot, 1983; Purkey and Smith, 1983).

The committee's recommendations for indicators of the effectiveness of science and mathematics teaching are based on this conception of the professional teacher and the support that the schools must provide to attract and retain such teachers. The rest of the chapter is organized into three categories of information about professionalism in science and mathematics teaching:

1. What are the educational backgrounds and knowledge levels of individuals who teach science and mathematics?
2. How do these individuals spend their time?
3. What are the working conditions for teachers of science and mathematics?

EDUCATIONAL BACKGROUNDS AND LEVELS OF KNOWLEDGE

College Education

One attribute of professional teachers is that they understand the subjects that they teach. To assess the extent to which the nation's secondary school science and mathematics teachers have adequate subject matter preparation, NSF has sponsored two surveys that have collected information on the education that teachers received in college in the subject matter fields that they teach (Weiss, 1978; Research Triangle Institute, 1985). More than half the states also collect information on college courses in science and mathematics taken by newly hired teachers. At present, the Center for Education Statistics of the U.S. Department of Education is considering plans for collecting information on teachers' undergraduate major and minor fields of preparation and, for both secondary and elementary teachers, on the number of college courses taken in mathematics and science and in teaching mathematics and/or science (Darling-Hammond et al., 1986). The premise underlying this sort of survey is that high school physics teachers, for example, who have taken little physics in college are unlikely to have a solid understanding of physics and consequently are unlikely to have the knowledge needed to teach physics well.

The committee recognizes that the extraordinary variety of undergraduate institutions in the United States that prepare teachers makes it virtually impossible to assess accurately the subject matter preparation of the nation's teachers. Nonetheless, we support continuation of the collection of information on teachers' college courses and degrees because it will provide at least basic information on the preparation of the teachers who teach science and mathematics to different types of children in the United States. Moreover, information on changes over time in teacher preparation and in the distribution of teachers among diffferent types of students will provide a sense of direction about the nation's success in staffing all schools with teachers who are well prepared in science and mathematics.

The committee does suggest one major change in the data collection and reporting method: information on teacher preparation should be collected and reported according to different subgroups of students taking mathematics and science courses, so that the information will be more useful in assessing the distribution of well-prepared teachers among groups of students with different characteristics. For this purpose, data collected on individual students should include gender, race, ethnicity, socioeconomic status, grade level, type of community (urban, suburban, rural), and region or state. Reporting by student subgroups will allow the following types of questions to be addressed:

- What proportion of the students taking high school physics are taught by teachers who have an undergraduate major or minor in physics?
- Is the proportion of black students studying biology with teachers who have an undergraduate major or minor in biology different from the percentage of white students studying biology with teachers with the same preparation?
- What proportion of elementary school students in particular grades and with particular characteristics are taught science by a teacher who has taken at least six college courses in science? What proportion are taught mathematics by a teacher who has taken at least six college courses in mathematics?

As the questions indicate, collecting and reporting information on teacher preparation by student subgroups permits one to examine whether the college education of the teachers who teach science and mathematics to students with particular characteristics differs from the college education of teachers teaching children with other characteristics. This strategy supports the focus on equity and access that the committee endorses. It will make it possible to learn whether the teachers with the most substantive college backgrounds are being selected to teach certain categories of students rather than others. This strategy also reduces the problem of how to assess the subject-matter knowledge of teachers teaching both mathematics and science and of high school teachers who teach more than one type of science.

Subject-Matter Knowledge

It has proven very difficult to establish that teachers with superior subject-matter knowledge are more effective in teaching students than are teachers who have merely an adequate knowledge of the material they teach to students (Byrne, 1983). For example, the evidence on the relation between graduate credits or advanced degrees and effectiveness is tenuous (Begle, 1979; Shymanski et al., 1983;

U.S. General Accounting Office, 1984). Nevertheless, it is reasonable to believe that teachers who have mastered the material that they teach to their students are more effective than teachers who have not mastered this material. Therefore, some appropriate measure of subject-matter knowledge should be used as an indicator of teacher effectiveness, even though agreement on specifics of optimal preparation for teaching a subject at a given grade level or in a particular course remains difficult. For this reason, the committee endorses periodic sample testing of teachers' basic competency in the subject matter they teach. The problem to date has been the development of an appropriate measure. Even if the relationship between subject-matter knowledge and effective teaching of a subject were better understood, there would still be problems with current tests analogous to those discussed in Chapter 4 with respect to tests of student learning. The committee suggests that the tests used to establish basic subject-matter competency of teachers should probe essentially the same domain as the tests used to assess students' mastery of science and mathematics. The results of this testing should be reported in summary statistical distributions rather than as individually identifiable scores, since the purpose is to establish an indicator of teachers' knowledge of the subject matter being taught, not to evaluate individuals in order to make decisions on hiring, promotions, or pay.

In implementing the committee's recommendation to test teachers' basic subject-matter competence in science and mathematics, it will be important to retain linkages not only to changes in the disciplines themselves but also to changes in science and mathematics curricula and in the content and form of student tests. In Chapter 4, the committee recommends that new tests be designed that more adequately assess students' higher-order thinking skills than existing tests and that are more closely tied to exemplary curricula. As the tests used to assess students' science and mathematics knowledge and skills change, so should the tests used to assess teachers' basic subject-matter competence. In this manner, any systematic deficiencies can be uncovered in teachers' mastery of the changing material on which their students are being assessed.

As with subject-matter preparation and for the same reason, the results of the teacher tests should be reported by student subgroup. Reporting the percentage of students with particular characteristics who are taught mathematics or science by teachers who possess basic subject-matter competence supports the committee's desired focus on equity concerns in the development of useful indicators of the quality of science and mathematics education.

Clearly, a measure of teachers' mastery of the same knowledge and skills on which their students are tested provides only a modest amount of information about their subject-matter competence. Even so, results of such tests may show that not all teachers have mastered the basic knowledge and skills. It is important to recognize that not all the reasons one might posit for this possible outcome blame teachers. For one thing, school district responses to declining enrollments during the 1970s—and in some parts of the country, during the 1980s—led to many teachers being reassigned from such fields as history or social science, in which there was a surplus of teachers, to such fields as mathematics and the physical sciences, in which there were vacancies (Darling-Hammond, 1984; Flowers, 1984). Often the preparation of these teachers in science or mathematics was very limited and outdated. Unfortunately, not all mandated changes in curriculum or in skill emphasis are accompanied by adequate inservice programs for the teachers who are required to implement the new ideas. Future shortages of qualified mathematics and science teachers may continue to induce some school districts to staff science and mathematics courses with teachers with little preparation or knowledge in these subject areas.

It is important to reiterate that the reason we recommend testing teachers' basic subject-matter competency in science and mathematics is to assess the extent to which students are taught science and mathematics by teachers who have mastered the knowledge and skills they teach, not to denigrate the ability or aptitudes of particular teachers. Thus, although the committee advocates collecting information on the characteristics and backgrounds of the students who are taught by the teachers sampled for testing, we do not suggest that comparable data be collected on the individual teachers being tested. However, since the ultimate goal is to provide teachers competent in science and mathematics to all students, individual states may wish to collect demographic data on teachers in order to examine the question of whether out-of-field teacher placement, access to in-service opportunities, and high-quality teacher preparation programs are evenly distributed among different population groups of teachers.

TABLE 6-1 Suggested Schedule for Assessing Subject-Matter Knowledge of Teachers of Science and Mathematics

Year	Teacher Survey (elem. and sec.)	New Hires (sec.) Survey	New Hires (sec.) Follow-Up
1988	X		
1989			X
1990		X	
1991			X
1992	X		
1993			X
1994		X	
1995			X
1996	X		

Sampling Strategy

The subject-matter preparation and subject-matter knowledge of a random sample of the nation's science and mathematics teachers ought to be assessed at least every four years. The sample should be drawn so that it is possible to discern trends not only in the preparation and subject-matter knowledge of the nation's science and mathematics teachers as a whole, but also trends in the preparation and knowledge of such critical subsets of teachers as those teaching particular sciences, those teaching remedial mathematics, those teaching science in the elementary schools, those teaching minority group children, and those teaching special education students.

In addition, the subject-matter preparation and subject-matter knowledge of a sample of newly hired secondary school science and mathematics teachers should be assessed every two years, with a follow-up survey administered one year after the original survey to determine whether the new hires are still teaching and, if not, why they left teaching (see Table 6-1 for suggested survey schedule). Newly hired teachers in this context are defined as those teachers employed to teach mathematics or science within the last year who did not teach mathematics or science in the year prior to this employment.

There are three reasons to focus particular attention on newly hired teachers. First, collecting information every two years on the college backgrounds and subject-matter knowledge levels of this group is one way to provide early warning of incipient changes in the backgrounds and skills of the profession. Second, the newly hired teachers are the most likely to leave teaching (Charters, 1970; Greenberg and McCall, 1974; Murnane, 1981). By learning which newly hired teachers leave teaching after one year, it is possible to examine whether those who leave have better preparation and subject-matter knowledge than those who stay, as one study of North Carolina teachers has found (Schlechty and Vance, 1983). Moreover, by learning what teachers who left did in the year after they left, it may be possible to make inferences concerning whether changes in salaries or working conditions might have induced these teachers to stay in the classroom. A third reason to study newly hired teachers is that the resulting information could shed light on sources of supply of new teachers in mathematics and science. For example, recent studies of newly hired science and mathematics teachers in Connecticut (Connecticut State Department of Education, 1985), Illinois (Illinois State Board of Education, 1983), and New York (New York State Education Department, 1983) indicate that the majority were teachers with previous teaching experience—members of the much discussed but elusive "reserve pool" of individuals who are certified to teach but are not currently employed by any school system. Very little is known about the size of the reserve pool or about the backgrounds and skills of individuals in this pool. In fact, the U.S. Department of Education's current model for national teacher supply and demand does not even acknowledge the reserve pool as a source of supply (Panel on Statistics on Supply and Demand for Precollege Science and Mathematics Teachers, 1987).

By collecting information biennially on the backgrounds and subject-matter knowledge of newly hired mathematics and science teachers and determining whether these teachers are still in the classroom in the next year, it would be possible to learn:

- whether the reserve pool is a greater source of supply of science and mathematics teachers in some parts of the country than others;
- whether the significance of the reserve pool as a source of supply of science and mathematics teachers changes over time;
- whether the educational backgrounds and knowledge levels of the newly hired coming from the reserve pool differ from the educational backgrounds and skills of the newly hired coming directly from teacher education programs; and
- whether the newly hired teachers coming from the reserve pool are more or less likely to remain in the classroom than those coming directly from teacher education programs.

Needed Research*

More research is needed on the impact that teachers' knowledge of science and mathematics has on their effectiveness in teaching these subjects to students. Current findings give no clear indication on the optimal breadth and depth of preparation for a given teaching assignment, whether an elementary grade or a secondary school course. Results of studies attempting to relate measures of teacher knowledge to measures of student achievement have been mixed (Summers and Wolfe, 1977; Begle, 1979; Byrne, 1983; Druva and Anderson, 1983). It is particularly important to learn whether new directions in curricula, especially increased attention to the development of students' analytical and critical thinking skills, will increase the importance of teachers' subject-matter knowledge. To address this question, teachers' effectiveness must be measured by students' scores on tests that assess these higher-order thinking skills and are closely tied to curricula. As noted above, such improved tests of students' skills should inform the tests that are used to measure teachers' basic subject-matter competence.

It would also be valuable to learn more about the roles that early home and school experiences play in determining the interest of elementary school teachers in science and the time they spend teaching science to their students. It would also be useful to learn whether early home and school experiences influence the decisions of college students to become science or mathematics teachers, and how long individuals who do start to teach science or mathematics remain in the teaching profession. Current evidence is mixed concerning whether there is greater attrition among science and mathematics teachers than among teachers in general (Cavin, 1986; Murnane, in press).

Some potential for such studies already exists, using data from the National Longitudinal Study of the High School Class of 1972 (NLS72) and the data base of the High School and Beyond Survey. A special supplement to the NLS72 fifth follow-up study, which was administered in 1986 to those members of the original sample who trained to become teachers, will increase the potential for such studies. The possibilities will be further enhanced during the 1990s by the National Educational Longitudinal Study (NELS), which will follow a large sample of students from their eighth school year in

---

* This section draws on the material presented in Haggstrom et al. (1986).

1988 through further schooling and into the labor market. Research should be supported that will use these and other data sets to explore the impact of early school and home experiences on the career choices of potential teachers.

Recommendations

Key Indicators: The committee recommends that samples of teachers be selected to take tests that probe the same content and skills that their students are expected to master. For this purpose, tests for teachers should be developed to include the same kinds of improvements that the committee recommends for tests of student learning.

The distribution of teachers' test scores should be reported by student background and characteristics (race, ethnicity, gender, socioeconomic status, type of community—urban, suburban, rural). This will provide information about the distribution across different student subgroups of teachers who are in command of the mathematics and science they are expected to teach. Both current distribution and change over time are of interest; therefore, tests should be given every four years to a sample of all teachers and every two years to a sample of newly hired secondary school mathematics and science teachers.

Supplementary Indicator: The committee recommends reorganization of the information currently being collected on teacher preparation (college courses in mathematics and science, majors and minors, advanced degrees), using the various student groups taught as the reporting groups of interest.

The information reported should display the percentage of students with particular backgrounds and characteristics who are being taught mathematics and science in elementary school as well as courses in these domains in secondary school by teachers with specific college preparation. For this indicator also, four-year cycles are appropriate for collection and analysis of information.

Research and Development: The committee recommends that research should be undertaken on two issues: the impact of teachers' knowledge of subject-matter on their effectiveness in teaching these subjects to students, and the role of early home and school experiences in determining decisions to become a teacher and on how and what to teach.

TEACHERS' USE OF TIME

Among the most significant professional decisions that teachers of mathematics and science make is how to spend their time. These decisions influence the skills, energy levels, and experiences that teachers share with their students. For this reason, the committee wishes to focus attention on collecting information in two relevant categories: time-use outside the classroom, both during and beyond the school day, and time-use within the classroom.

Time-Use Outside the Classroom

One of the attributes of professional teachers is that they continue to learn as they teach and continue to evince interest in the subjects they teach. Thus, one question that should be explored is whether schools are making progress in attracting and retaining teachers of science and mathematics who spend some of their time outside the classroom in activities that demonstrate their continued intellectual curiosity about the subjects they teach. A second question related to time-use is whether school policies are changing the way in which teachers of mathematics and science at various levels are spending their time in school when not actually teaching in the classroom. Teachers who have no time to develop collegial relationships or to plan mathematics and science activities will be less able to exhibit characteristics of fully professional teachers (Rosenholtz, 1985). A third question is whether the salary increases that are taking place in many states are making it possible for more teachers to live on their teaching salaries, lessening the pressure to seek second jobs unrelated to teaching that reduce the time and energy these teachers can devote to their profession. (Currently about 10 percent of teachers earn additional income in jobs outside the school system; 20 percent earn additional pay within the school system for nonclassroom functions; [National Center for Education Statistics, 1985].)

A difficulty involved in collecting information on teachers' time-use is the diversity of ways that teachers can engage in intellectually rewarding pursuits—far more ways than could be anticipated in a closed-ended survey instrument. Similarly, it may be difficult to anticipate in a closed-ended questionnaire the number of tasks or assignments that could inhibit a teacher's ability to use time in professional ways. For this reason, we suggest that information on science and mathematics teachers' time-use be collected every fourth year through a time-budget study. In this approach, a sample of teachers of mathematics and science at different grade levels would be asked to keep a diary, recording how they spend their time during a particular period, perhaps a week. The study should be structured so that data are collected on teachers' time-use during different parts of the year.

While time-budget studies have little precedent as a data collection strategy in education, they have been used extensively and informatively in research in other areas. Research based on data from time-budget studies has revealed important, and unsuspected, patterns in how families (Juster and Stafford, 1985), children (Medrich et al., 1982), and college professors (Institute for Research in Social Behavior, 1984) spend their time. Analogous research on teachers' use of time would be particularly valuable at a time when states and local school districts are engaging in myriad activities that change the incentives and constraints influencing teachers' behaviors—salary increases, career-ladder plans, periodic testing of teachers' subject-matter knowledge, and more intensive evaluations of teachers' in-class performance (Goertz et al., 1984; Goertz, 1986). There are many ways in which teachers could respond to these new incentives and constraints, and some of the responses could be quite different from those intended. Learning about changes in how teachers of science and mathematics spend their time outside the classroom will throw light on teachers' responses to the many policy changes aimed at improving the quality of education in these subjects.

The advantage of the open-ended nature of the time-budget approach is that little prior categorization need be imposed—that is, it is not necessary to prepare a list of how teachers might spend their time before beginning the study. In analyzing the data generated by any such time-use study, however, some assumptions will have to be made as to which activities contribute to a teacher's professional competency. Candidate activities include engagement in professional association activities, work on publications related to science or mathematics, graduate study on the use of computers, time spent in preparation of lessons and new courses, work on curriculum development, professional relations with colleagues, and working with students beyond classroom hours. In addition, time-use might be identified that might contribute negatively to a teacher's energy and enthusiasm for teaching science and mathematics. Analyses of data generated in time-use studies will reveal interesting patterns and provide suggestions for future indicators with specific policy implications. The establishment of trends over time will be important for assessing the effects of policy changes.

Time-Use in the Classroom

The use of classroom time involves both teacher and student behavior. At the elementary level, teachers have a great deal of control over student behavior in that they control instructional time. Thus, when their teachers allocate more classroom time to teaching, say, mathematics, elementary school students learn more mathematics (Wiley and Harnischfeger, 1974). The committee is concerned by evidence indicating that some elementary school teachers devote almost no classroom time to science instruction (Weiss, 1978). For this reason, we reiterate here the recommendation made in Chapter 5 that data be collected every four years on the percentage of elementary students whose teachers devote at least a minimal amount of classroom time to science instruction. This minimal level needs to be chosen with care and should reflect the time needed to teach a meaningful science curriculum at different grade levels. As noted in Chapter 5, student groups with various characteristics should be the unit of analysis in this data collection, and the information should be reported in terms of the percentage of students with particular characteristics and backgrounds in different grade levels who spend at least so many minutes of the school instructional day on science. It will be particularly important to observe whether changes take place over time in the reported distributions as state policies mandating greater attention to science instruction are put into place. The discussion here elaborating the recommendation in Chapter 5 with respect to minimal time is framed in terms of elementary school science instruction and puts less emphasis on time devoted to mathematics instruction because it is the committee's sense that mathematics is treated as a major subject by most elementary school teachers. Nevertheless, the variability in time spent on mathematics instruction in different grades (Romberg and Carpenter, 1986) makes this measure important for both subjects.

Homework also entails both teacher and student behavior. Students learn more when they do more homework in mathematics and science (Walberg et al., 1986), and homework is more effective when it is relevant to the student learning desired, teachers check it regularly, and students are given feedback about the quality of the homework they complete (Walberg and Rasher, 1986). Within these qualifications, the committee in Chapter 5 recommends the collection of data every four years on the fraction of students with particular characteristics and backgrounds who are regularly assigned effective homework by their teachers. If desired, the benchmarks for assessments of desirable quantities of homework may be determined by expert panels, although both geographic (e.g., international) and temporal comparisons are in themselves useful. For example, changes over time in the reported distributions will provide useful information about one important predictor of students' science and mathematics achievement.

Clearly, how much science and mathematics students learn in school depends not only on the amount of time devoted to science and mathematics instruction and on the amount of homework assigned, completed, and corrected, but also on how classroom time devoted to science and mathematics is used. While the committee has recommended observation of student behavior in this respect (see Chapter 5), we do not recommend the development of indicators that assess teachers' use of classroom time. The reason is that our interpretation of the process-product and teacher effectiveness literature leads us to the aforementioned conclusion that effective teaching requires the orchestration of a variety of strategies suited to the specific instructional context and therefore cannot be characterized by the routine use of particular well-defined actions. Consequently, it would not be useful to base indicators on counts of how often teachers engage in such actions.

Needed Research

Recent state and local initiatives are making important changes in the intended science and mathematics curricula. How these changes influence the curriculum content that students actually encounter depends to a large extent on teachers' responses to the changes in the intended curricula. Past research indicates that these responses can vary, depending on the subject-matter, the skills of the teachers, the adequacy of in-service education programs, the availability of facilities and materials, and the attitudes of administrators (Berman and McLaughlin, 1974–1975; Sarason, 1985). Learning more about the factors that influence teachers' responses to changes in the intended curriculum will require expensive, painstaking observational research. The research is needed, however, to understand how state and local curricular initiatives affect the quality of science and mathematics instruction provided to students.

In the committee's view, there are two dimensions of teachers' use of time in the classroom that may merit significant research attention, possibly because they might one day lead to new indicators, but more significantly because they might improve the quality of science and mathematics instruction. The first of these dimensions is the use of concrete materials, laboratory experiments, and computers in the classroom. Such hands-on components of the curriculum may be particularly important in the development of higher-order thinking skills. Consequently, in evaluating the effectiveness of hands-on instruction, it is important to use student tests that measure higher-order skills.

A second dimension of teachers' use of time involves questioning techniques. There is evidence that students' learning increases when teachers wait at least five seconds for student answers to questions (Rowe, 1983). Learning more about the accomplishments and limitations of increases in waiting time may provide the basis for an indicator of one dimension of effective teaching. More important, such research may result in improvement in the quality of science and mathematics teaching in the schools.

Recommendations

Supplementary Indicator: The committee recommends that time-budget studies be conducted, asking teachers to record how they spend time related professionally to their present or future classroom activities, other than in the classroom itself, during a particular period, perhaps a week.

The information collected should be evaluated against sets of activities identified by experts as advancing or hindering effectiveness in the classroom in teaching mathematics or science. Investigations of the relationships between professional activities reported by teachers and teaching effectiveness should be conducted to help refine this indicator.

Research and Development: The committee recommends research on the following aspects of the behavior of teachers in science and mathematics instruction (see also the related research recommendations in Chapter 5 on student behavior):

- the factors affecting teacher responses to changes in the intended curriculum;
- the use of hands-on experiences involving concrete materials, laboratory experiments, and computers; and
- allowing an adequate period of time for students to formulate responses to questions.

The recommendations in Chapter 5 on the amount of time given to the study of science and mathematics in elementary school and on the amount of homework can be considered indicators of teacher behavior as well as student behavior. In either case, we consider them important indicators of the quality of science and mathematics education.

IMPLICATIONS FOR STATE EDUCATION AGENCIES

Up to this point, the emphasis in implementing teacher evaluation schemes in the various states has been on knowledge of the subject matter rather than on other characteristics. A major exception is Tennessee, which more comprehensively than other states has developed an on-site observation and interview schedule to complement simple subject-matter knowledge. This approach needs to be more fully explored if a more complete picture of science and mathematics education is to be drawn.

The main data source currently available to states for analyzing teacher effectiveness is subject-matter knowledge of teacher candidates. What is not known (because it is not systematically analyzed) includes the following:

- Are there significant variations among objectives that all new mathematics and science high school teachers as well as elementary teachers need to know, as reflected in teacher job-analysis surveys, polls of college of education faculty, test questions, and test results?

Are the variations greater from state to state, between school systems of different types (e.g., large urban versus rural) within a state, between different sorts of institutions preparing teachers? Is a national consensus emerging on what individuals need to know to be effective science or mathematics teachers?

- Is science knowledge part of the requirements for elementary teachers? Tests for elementary teachers generally lack science content; typically, they are dominated by questions on general pedagogy. The low expectation for instruction in science at the elementary level may be a contributing factor, as may be the absence of any agreement as to what the science content of the elementary school curriculum should be, even when science is being taught.
- With regard to testing for certification: Are there fewer minorities, proportionately or in actual numbers, entering teacher-preparation programs than in the past—especially those training to be future mathematics and science teachers? Are tests and test results such that they systematically discourage members of some population subgroups from choosing teaching careers? Are there patterns in geographic distribution of the lowest-scoring test takers—for example, are they entering urban schools or small rural ones in greater proportion than suburban schools?

The periodic collection and analysis of even this small part of the information needed about the potential education work force could have the following state-level policy implications:

- Recruiting and preparing minority teacher candidates may need to begin in the junior year of high school; special scholarship programs may have to be initiated especially in mathematics and science if the number of minorities in these fields falls significantly below a predetermined standard.
- Approval procedures for undergraduate teacher-education programs could be revised to ensure that prospective teachers are exposed to sufficient mathematics and science experiences.
- Entrance examination systems for teacher-education programs may need to be structured in such a way as to provide diagnostic information about the strengths and weaknesses in mathematics and science of entering candidates; such profiles could be used to guide candidates to specific academic sequences that would ensure that they had at least been exposed to appropriate mathematics and science courses.

- Analysis of the mathematics and science test results from successful teacher candidates could lead to targeted regional and state staff-development programs if it is found that the least prepared teachers are locating in certain areas.
- If the committee's recommendation to follow up the candidates who pass the certification tests and become new hires were to be implemented, it could establish a useful data base on the ability of the education establishment to provide conditions that induce teachers to stay, thereby assisting in future projections of supply and demand.

The precertification information can be collected and used annually in those states that possess the requisite data base. However, few states carry out systematic testing of certified teachers, and it is unlikely that this approach will become more widespread. Even if it did, the results would not enrich the general knowledge about teachers because current tests typically avoid science and touch only the basics of mathematics.

The committee's recommendation on teacher testing rejects any connection between the use of a nationwide sampling of teachers' mathematics or science knowledge and any use of the information for purposes of personnel decisions. Instead, the data from the tests recommended by the committee would provide a national benchmark on the continuing intellectual growth of school faculty and whether they are staying current. Such data would provide to the states as well as other units of government information that could drive the creation of relevant staff development programs and materials.

Finally, assuming some consensus within a state on curriculum, observation of how teachers of mathematics and science organize and present the material and the context in which they present it (time spent on planning and presenting, availability of equipment, etc.) become important indicators of teaching quality in a state's schools. This is especially so since more state legislative bodies are requiring local as well as state "report cards" to document class time spent in subject areas. By themselves, the statistics on minutes spent per day or week on a curriculum area are almost meaningless; they can become indicators only in conjunction with information on other variables. For example, collecting information on whether pupils are asked weekly to write a 250-word science laboratory report is quite superficial; it takes on meaning only when one also knows how often these same reports are actually read and critically evaluated, with the results returned to the student. Only then is the writing requirement likely to help improve the quality of student understanding of science.

WORKING CONDITIONS FOR TEACHING SCIENCE AND MATHEMATICS

Resources for Teaching Science and Mathematics

Effective teaching is best sustained if schools are places where professional teachers like to work and places that provide support for activities that characterize effective teaching. Consequently, it is important to develop indicators of the extent to which the nation's schools are able to provide the resources and support needed to sustain fully professional teaching of science and mathematics for all students. For reasons explained in Chapter 8, the committee does not recommend the collection of data on per-pupil expenditures devoted to science and mathematics or on specific budgets available to science and mathematics teachers. What we do see as important, however, is to collect detailed information on the uses to which money devoted to mathematics and science instruction is put within a school and within a classroom.

The following information on working conditions in schools is pertinent:

- the availability and use of equipment, materials, textbooks, and laboratory facilities appropriate to the intended curriculum;
- the number of students and different types of courses taught by each teacher;
- the availability and use of professional time for planning during school hours, and support for professional activities (further education, curriculum development, collegial exchanges) during the year and during summers; and
- the availability and use of assistance such as classroom or laboratory aides.

At first glance, this information may appear relatively easy to collect using closed-ended questionnaires. This may not be the case, however, for several reasons. First, the mere presence of a facility or materials and equipment does not ensure their use. Even in 1965, most secondary schools had, for example, some facility that was called a laboratory (Coleman et al., 1966). Analyses of the data indicated relatively minor differences among schools in the number of facilities. Most analysts believe, however, that in 1965 and in 1987 as well, there were and are significant differences in the quality of the equipment, materials, and laboratory facilities present in different schools. It is very difficult to capture these differences in quality with closed-ended survey instruments. In addition, as school district officials pointed out (see Appendix C), a secondary school may have adequate laboratory facilities, but only students taking advanced science courses may have access to them. An elementary school may have a few classrooms with provisions for hands-on work, but these may not be available to all grade levels or all classes at a single grade level. It is difficult to learn from closed-ended surveys the extent to which all students taking science and mathematics have access to a school's equipment, materials, and laboratory facilities.

Similarly, materials and equipment may be present in a school, but the procedures for making use of them may be so bureaucratic that teachers forego the opportunity to use the potentially available equipment and supplies in their teaching. This suggests the importance of learning about teachers' control of equipment and supplies, and whether teachers actually employ the equipment, supplies, and laboratory facilities in their teaching.

For these reasons, we suggest that pilot studies be conducted to explore whether a macro-level indicator can be developed using information on the conditions under which teachers of science and mathematics work. The information should be collected through the use of open-ended interviews. All teachers and administrators who are interviewed would be asked the same questions, with special attention to probing teachers' open-ended answers. While it will be more difficult to organize these open-ended responses than it would be to tabulate teachers' responses to closed-ended questionnaire items, we consider the open-ended questionnaires to be a much more effective strategy for gathering reliable information about the conditions under which science and mathematics teachers work, the number of students taught under inadequate conditions, and changes over time in teachers' access to the resources needed to do their job well. If pilot studies indicate the feasibility of developing an indicator on resource use and working conditions, the information should be collected every four years. Such an indicator, as other indicators described in this chapter and elsewhere, should be expressed in terms of percentages of students of different backgrounds and characteristics who are being served. Careful attention will have to be given to sample design to achieve comparability over time as well as generalizability.

Salaries as Incentives

Teacher salaries tend to rank relatively low among professional salaries. This may discourage individuals from entering or staying in teaching, particularly those with training in mathematics and the physical sciences who may have attractive, alternative career opportunities.

Even if potential teachers' career decisions were not sensitive to the financial rewards in teaching relative to those in other professions, it would still be somewhat anomalous to pay poorly the members of a profession who potentially can have such marked effects on children's futures. Nevertheless, one might argue to retain the low pay for financial reasons if it did not affect the decisions that teachers and potential teachers make. There is strong evidence, however, that teachers' decisions are influenced by salaries. For example, Freeman (1976) and Zarkin (1985) have shown that the number of college students who study to become teachers is very sensitive to relative salaries. In addition, Manski (1985) found that the number of academically talented college students who enter teaching is affected by salaries. Subsequent career decisions are also influenced by salaries, for example, teachers' decisions to move from one school district to another and their decisions on whether to leave teaching entirely (Eberts and Stone, 1984). Thus, salaries appear to provide incentives that have measurable impacts on the career decisions of teachers and prospective teachers and consequently influence the ability of the nation's school districts to staff schools with competent teachers.

Since salaries in business and industry vary by subject-matter field, comparative salary data need to be collected by field of specialization. This is illustrated by Figure 6-1, which displays data on average starting salaries in business and industry, expressed in 1967 constant dollars, for college graduates with bachelor's degrees in particular subjects. These data are derived from surveys administered by the College Placement Council. For the purpose of comparison, Figure 3 also displays data on average starting salaries for elementary and secondary school teachers expressed in 1967 dollars. In interpreting the teachers' starting salary data, which stem from surveys administered by the National Education Association (NEA), it is important to keep in mind that more than 99 percent of U.S. public school teachers work in school districts using uniform salary scales, under which field of specialization has no effect on salary. As a result, in any given district, the starting salary of a physics teacher is the same as the starting salary of a history teacher.

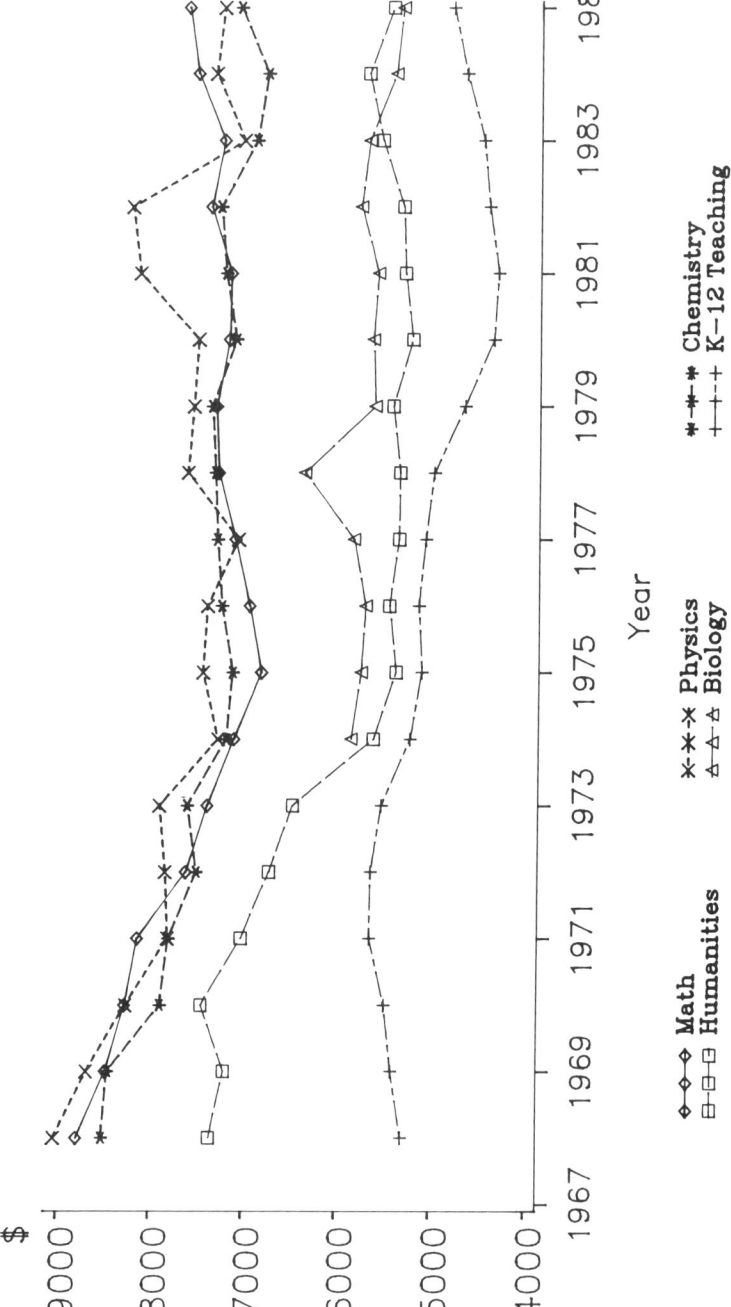

FIGURE 6-1 Starting salaries in 1967 dollars for inexperienced bachelors candidates by degree specialization. Sources: College Placement Council; National Educational Association.

INDICATORS OF TEACHING QUALITY 115

Figure 6-1 illustrates two points. First, how much more a college graduate earned by taking a job in business or industry than by taking a teaching position depends on the graduate's subject speciality. For example, in 1974, college graduates specializing in mathematics, chemistry, or physics who entered business or industry were paid 36 to 39 percent more on average than college graduates who became teachers, while college graduates trained in the humanities who entered business or industry were paid only 7 percent more on average than college graduates who became teachers; for graduates trained in biology, the differential was 12 percent.

Second, the salary differentials between business and industry and teaching have changed over time, and the pattern varies among subject specialties. In general, the differential between teaching and other occupational alternatives has increased more for graduates trained in mathematics or the physical sciences than for graduates trained in the humanities or biology. For example, in 1985, the starting salary advantage that business and industry offered over teaching had risen to 59 percent for graduates trained in mathematics, but it had risen to only 13 percent for graduates trained in one of the humanities, and had actually fallen by one percentage point for biology graduates. These data indicate the importance of considering each field separately.

Comparative salary data need to be collected every two or three years because salaries in different occupations can change significantly from year to year, and changes over time in the salaries offered in different occupations are more informative than salary comparisons at one point in time. In fact, it is not possible to judge from comparisons of starting salaries at one point in time whether the schools are able to attract talented college graduates into teaching. One reason is that working conditions may differ between jobs in teaching and jobs in business or industry. A second reason is that the comparative salary figures are very sensitive to the method of calculation. For example, when daily salaries are compared by dividing annual salaries by number of required work days (180 to 200 for teachers; 240 for college graduates working in business or industry), teachers' salaries appear more attractive than when annual salaries are compared. There is no one right way to do the calculation: teachers' work days during the school year may be very long days (the proposed time-budget study would address this issue), and many teachers do not have work opportunities during the summer at the same rate of pay. In contrast to the difficulty of making inferences from comparative salaries at one point in time, trends in comparative salaries do provide important information about changes in the ability of the schools to attract talented college graduates with particular types of training.

The following salary data should be collected at least every three years (preferably every two years) for each field of study (for example, mathematics, biology, physics, chemistry): (a) information on starting salaries in teaching and in business and industry and (b) information on salaries after 15 years of experience. The latter information is important because, in choosing fields of specialization and occupation, college students do compare not only starting salaries, but also streams of earnings (Zabalza et al., 1979). Moreover, differences in starting salaries between occupations do not always reflect differences in salary streams. For example, the average salary advantage of industry over secondary school teaching was 49 percent ($32,100 compared with $21,600) for individuals with 0 to 4 years of work experience after earning a master's degree in physics; the differential was 70 percent ($50,300 compared with $29,500) for individuals with 15 to 19 years of work experience after earning a master's degree (American Institute of Physics, 1983). The information on starting salaries and on salaries after 15 years of experience should include median salaries and the interquartile range of salaries. Median salaries provide a measure of central tendency—an indicator of what the average person in a particular occupation with a particular amount of experience earns, while the interquartile range reflects the amount of variation, for example, in the earnings of a particular group. A large interquartile range may make a particular occupation less attractive, in that college students cannot count on receiving a particular level of compensation if they choose that occupation.

There are important differences between the committee's proposals for salary comparisons and comparisons of average salaries in different occupations. The latter comparisons, which are often cited in the media, can be deceiving because they are sensitive to the distribution of experience in each occupation. For example, average salaries in teaching grew more rapidly during the 1970s than starting salaries did because the teaching force became older during the decade, since relatively few new teachers were hired. Thus, average salaries do not necessarily reflect the attractiveness of teaching to college graduates who are making occupational choices.

The salary comparisons proposed by the committee will throw

the most light on the competitiveness of secondary school teaching salaries, at least for the present, since it is mainly secondary school teachers who have college majors in the subjects that they teach. This may be changing, however, as some states and institutions of higher education follow current proposals to eliminate undergraduate degree programs in elementary school education.

Developing the suggested indicator of salary differentials can take advantage of a number of already existing salary surveys. For example, the College Placement Council collects data annually on the salary offers made to a sample of college graduates with particular subject-matter specialties. For many years, the Northwestern Endicott Report (1985) has provided annual information on the salaries that a sample of large business and industrial concerns pay to college graduates with particular subject-matter specialties. The U.S. Department of Labor also makes available biennial reports of starting salaries in private industry for college graduates with certain specialties. Several professional associations, including the American Chemical Society, the American Institute of Physics, and the American Mathematical Society, publish annual reports of the average or median starting salaries earned by their members, broken down by highest degree earned (e.g., American Institute of Physics, 1983). Much of the salary information collected in individual surveys is presented in a biennial publication of the Commission on Professionals in Science and Technology (formerly, the Scientific Manpower Commission) entitled *Salaries of Scientists, Engineers, and Technicians*. The NEA, which is the primary source of data on starting salaries in teaching, does not routinely report average salaries for teachers with a bachelor's degree and 15 years of experience. However, the salary schedules that are used for the calculation of starting salaries would support generation of this information.

It would be preferable to have the data on comparative salaries generated by a single organization using one method. It is difficult to determine, for example, the extent to which differences in the median starting salaries of chemists and biologists reported by the respective professional societies stem from differences in survey method. One strategy that should be explored is the use of data from the U.S. Census Bureau's Current Population Survey to generate comparative salary data. Until a uniform method is developed, however, salary data can be reported using information generated by the sources cited above.

Recommendations

Supplementary Indicator: The committee recommends that data be collected on a four-year cycle through open-ended surveys on the materials, facilities, and supplies available and used by teachers in mathematics and science instruction.

An indicator can be constructed from this information by reporting on the levels of resources being used in the classroom by student subgroups of different backgrounds and competencies.

Key Indicator: The committee recommends collection at least every three years (preferably every two years) of detailed information on the salaries paid to college graduates with particular subject-matter specialties who choose to enter various occupations.

The information should include data on starting salaries and on salaries after 15 years of experience. These data should be reported in a manner that facilitates comparisons of salaries in teaching with salaries in other occupations for college graduates trained in particular sciences and mathematics.

7

Indicators of Curriculum

The curriculum interacts with teachers and students in complex and important ways. Classroom behavior is inseparable from curriculum. By providing incentives that stimulate effective teaching and learning, or by creating constraints on study and understanding, the curriculum affects the choices of students and teachers in every classroom. A curriculum may or may not provide incentives for teachers to master specific teaching techniques, such as laboratory experiments or the use of current events in creating applied mathematics problems. A curriculum might create opportunities for students to do extra work on questions raised in school, for example, by focusing attention on the evolutionary implications of insect species diversity. And a curriculum can act as a constraint on both teachers and students when the information conveyed through textbooks or tests is inaccurate, explanations are confusing or misleading, the logic of a concept and its derivation is lost, or mathematics or science is viewed as the memorization of facts and technical vocabulary.

These examples suggest the importance of the idea that a curriculum, by itself, does not cause teachers and students to behave in a certain way. Teachers or students can ignore a textbook, correct its errors, fail to carry out its inappropriate methods—and in so doing, create a learning experience that is better or worse, or simply different, from the one envisioned in the formal curriculum. But curricula still matter. By providing materials, encouragement, points of view, evaluations, and other pressures for certain approaches to teaching and learning, as well as discouragement and sanctions for others, curricula shape behavior. By portraying science or mathematics as it is actually practiced, or by substituting a dogmatic, rigid version, curricula signal to teachers and students how they are expected to behave if they continue their work in science and mathematics.

As important as curriculum is to the quality of science and mathematics education, no indicators exist to assess curriculum quality (Raizen and Jones, 1985). Science texts are reviewed from time to time by professional bodies, for example, by the American Association for the Advancement of Science (1985a, 1986a, 1986b), but this represents only a small slice of the curriculum and can address only partially the kinds of policy questions that confront teachers, educators, and others who need to make decisions about curricula.

DEFINING THE CONCEPTS

This chapter recommends the development of indicators to assess the coverage and quality of the mathematics and science curricula in the schools. Before considering how and why such indicators might be formed, two prior questions need to be addressed: What is meant by *curriculum*? Who will use curriculum indicators and how will they use them?

What Is Meant by Curriculum?

The curriculum is usually visualized as an operational plan that includes the substantive content, the expected actions and behaviors of teachers, the expected actions and behaviors of students, and the technology (textbooks, laboratory exercises, computer programs, tests, explicit pedagogic strategies) for conveying subject matter and structuring teacher and student activities. Indicators for two components of the curriculum thus broadly defined—the actions and behaviors of students and of teachers—are discussed in the two preceding chapters. Therefore, the term *curriculum* as used in this chapter (and generally throughout the report) refers primarily to the subject matter, the content of the curriculum. In mathematics and science, this includes theories, facts, algorithms, concepts, methods of inquiry, and procedural knowledge. Unless the text specifically notes otherwise, this definition is not concerned with much of the paraphernalia of curriculum that turns it into instructional chunks, including directions for sequencing and presenting the content. Although the committee recognizes that such instructional directions may also provide incentives and constraints, they ought to be matters determined as much as possible by the teacher, with whatever guidance is needed from other teachers and local and state curriculum specialists.

The substantive content of the curriculum generally represents a joining together of many different influences: historical precedent, views of professional educators, market forces determining the sales of textbooks and related instructional aids, the wishes of parents and other interest groups in the community, recommendations by state and national bodies, and changing perceptions of what students need to know. Moreover, the curriculum is expressed in several different forms: the plans and guidelines of state and local policy makers, the content of textbooks and such other materials as related workbooks and laboratory manuals, the actual content presented to the student, and the content learned by the student. These distinctions have been widely recognized. In this chapter, the committee, following the practice of IEA, refers to guidelines, textbooks, tests, and other written or programmed materials to be used for instruction as the *intended curriculum*; all of this material as constructed and presented by the teacher as the *actual* or *implemented curriculum*; and the content and skills learned by the student as the *achieved curriculum*. In the committee's view, it is important to have indicators of all three of these forms of the curriculum, since they would provide substantially different information and might be used to answer different policy questions.

The intended curriculum itself takes on many different expressions. In a concession to practicality, the committee decided to limit the scope of our recommendations to three manifestations of the intended curriculum: (1) the content of state plans and guidelines; (2) the content in textbooks and directly related workbooks, laboratory exercises, computer software, and other materials; and (3) the content of examinations. In a few states, for example, New York, state guidelines have always been an important determinant of the intended curriculum. As states become more active in school reform and assessment, state guidelines can be expected to play an important role in an increasing number of states. Regarding the second aspect of the intended curriculum, there is much research evidence that the content of textbooks importantly influences the content presented to the student (Goldstein, 1978; Stake and Easley, 1978;

Goodlad, 1984). Therefore, an indicator of the content of textbooks needs to be part of any monitoring system for science and mathematics education. As for the content of examinations, it also is believed to influence classroom instruction to a considerable degree (Resnick and Resnick, 1985; Romberg, 1986); hence, test content needs to be monitored as well. When the term *intended curriculum* is used alone in this chapter, it refers to any or all of these three levels of the formalized expression of curriculum, unless specific reference is made to a particular form, such as state guidelines.

In parallel with assessing the intended curriculum, it is necessary to assess the implemented curriculum, the curriculum that the student actually experiences. However, this is considerably more difficult. Whereas assessing the intended curriculum can be done by analyzing written materials apart from the classroom, assessing the implemented curriculum requires classroom observation. The third expression of the curriculum, the content learned by the student or the achieved curriculum, has already been considered in Chapter 4.

Another question regarding the meaning of the term *curriculum* concerns the grade-level span over which curricula are defined: Is it a school term, a grade in school, or a longer period of time, such as all the elementary grades? The committee believes that grade-level groups have greater validity for assessment than single grades because of the interrelated nature of much of the content of mathematics and science and because of the fact that there are many ways of teaching to reach productive educational goals. Prescriptions for attainment for each year of school would generate the kinds of lockstep curricula that constrain the creativity of teachers and are likely to lead to mediocrity. However, the committee believes that challenging, yet sensible, goals for defensible grade spans are critical for upgrading the general quality of mathematics and science education.

After some consideration, the committee suggests the following curriculum blocks as useful to consider as integral units: grades K–5, grades 6–8, the high school literacy curriculum, and the high school curriculum for college-bound students. The committee proposes that indicators for mathematics and science be developed for each of these grade clusters. Because the emphasis in this report is on mathematics and science literacy for all students, the committee views the need for assessing elementary and middle school curricula to be of the highest priority of the four areas, with the high school literacy curricula next in order of priority.

Some variations in grade clusters may be appropriate, for example, mathematics is often structured K-4, 5-8, 9-12, as in the forthcoming standards for school mathematics being prepared by the National Council of Teachers of Mathematics. In mathematics in particular, it may be of value to overlap the curriculum blocks so as to allow for greater flexibility of topic placement; blocks representing grades K-6, 5-9, and 9-12 have been suggested at recent international meetings of mathematicians and mathematics educators. There generally are two major options for the high school mathematics and science curricula for college-bound students: one for students expecting to major in mathematics and science-related fields and one for students expecting to major in other fields, another variation that should be considered in constructing the curriculum frameworks proposed below.

Indicators for Whom?

A second issue that concerned the committee had to do with the audience for the indicators of curriculum quality and their use by that audience. It is one thing to design a way of assessing the scientific quality of a textbook for a committee of scientists. It is another thing to design a way of capturing the quality of the science curriculum in a state for a state legislator with little science background. The committee concluded that the ultimate audience for its work should be federal, state, and local policy makers responsible for thinking about the overall quality of the educational program under their jurisdictions, even though specific judgments on the quality of the science or mathematics being taught will be made by scientists and mathematicians. Indicators should be developed to allow policy makers to address the following kinds of questions:

- How much attention is paid to complex problem solving by the schools in our state? Has this changed over time? Is it more or less than in other states?
- Do some kinds of children receive more mathematics content than others? By race? By social class? By sex?
- Do children in our schools receive as comprehensive an implemented science (or mathematics) curriculum as children in schools in Japan?
- What relationship do the state curriculum guidelines have to the actual instruction that goes on in the state?

- Have the recent state reforms in education changed the content and nature of science and mathematics education in our state and other states?

What Kinds of Indicators?

There is little question about the importance of the content of the implemented curriculum in determining the achievement of students in mathematics and science. At the most superficial level, it seems clear that few students would learn anything about geometry or the conservation of energy, for example, unless they received systematic instruction. One reason that Japanese 13-year-olds outscore their U.S. counterparts in mathematics is that all Japanese students are exposed to a year of algebra in seventh grade, while most U.S. students have to wait until ninth grade (McKnight et al., 1987). At a more detailed level, there is a solid literature that relates the teaching of particular concepts, knowledge, and skills to their acquisition by students (Walker and Schaffarzick, 1974; Wolf, 1977; Peterson, 1979; Romberg, 1986). IEA's second international mathematics study found that, in the United States, eighth-grade mathematics students are typically placed in one of four kinds of classes (remedial, typical, enriched, algebra). The amount covered is lowest in the remedial classes, about 25 percent greater in the typical classes, and another 25 percent greater again in the enriched classes. (Algebra classes were not included in this analysis because of their entirely different content.) The achievement gains of the students in the four types of classes correspond directly to the amount covered in the classes (Crosswhite et al., 1985). The extent of variability among curricula in content coverage, even given presumed variability in student ability, may well foreclose the possibility of attaining desirable levels of student achievement for some student populations (McKnight et al., 1987).

A description of the content coverage of a curriculum is only a beginning; in addition, descriptors of curriculum quality are needed. After all, topics can be included in a curriculum briefly or superficially. At one level, it seems obvious that students will have a better chance of learning something if sufficient time is allocated to learning it. This is the driving notion behind some instructional approaches, such as mastery learning (Bloom, 1976; Brophy and Good, 1986). Similarly, if a concept is introduced a number of different times during the school experience of a child, in different contexts and in increasing complexity, it is more likely to be well learned. This strategy leads to an approach—the spiraled curriculum—favored by many science curriculum specialists. (See, however, warnings by McKnight et al. [1987] against a poorly implemented spiral curriculum that can lead to shallow repetition of topics and attenuation of the curriculum.) At present, there is an emerging literature that relates the depth of coverage of subject matter to student understanding of the content (Glaser, 1984; Sizer, 1984). Deeper, more complex coverage of a concept or set of concepts increases the opportunity for students to be engaged in effective complex problem solving (Chi et al., 1981; Resnick, 1987). Not surprisingly, these researchers have also found that people's capacity to understand and remember new information in an area is related to their prior level of understanding of the area, and that experts in a field approach the solution of problems differently and more efficiently than do novices. This discussion suggests that the depth of coverage of material in a curriculum is an important aspect of its quality and needs to be assessed, in addition to the assessment of the extent of concept coverage.

The quality of a mathematics or science curriculum is influenced by two other factors: the scientific and mathematical accuracy of the content and the pedagogical logic or way it is presented. Curricula act as unwelcome constraints on the teacher's effectiveness to the extent that they embody inaccuracies, inadequate explanations, or poor sequencing of concepts or when they misrepresent the methods of science and mathematics, for example, by presenting scientific inquiry as a dogmatic and rigid procedure. No matter how comprehensive or deep the coverage of a content area, there will be little gained if it is confusing or inaccurate. Similarly, materials that are poorly organized or sequenced or exhibit other poor pedagogic strategies also constrain the teacher's ability to present the subject well. This suggests that an assessment of curricular quality needs to address the mathematical or scientific accuracy and the pedagogical quality of a curriculum, in addition to its depth.

To summarize, the committee suggests the development of two types of measures to capture and assess the content of mathematics and science curricula: measures of the extent of content covered in the curriculum and measures of quality including the depth of coverage of the content in the curriculum, the scientific or mathematical accuracy of the content of the curriculum, and the pedagogical quality of the curriculum. Measures of these two types should be developed for both the intended curriculum and the implemented curriculum.

Indicators to assess the kinds of policy questions set out earlier should be developed from these measures. For example, to assess breadth of coverage throughout a set of schools, the ratings of the different textbooks being used could be weighted by the number of students using each textbook.

MEASURES OF CURRICULUM CONTAINED IN OTHER CHAPTERS

There are clear relationships between the topics of concern treated in this chapter and several topics discussed in earlier chapters. One crude way of assessing coverage of content in high school, for example, is to measure the number of mathematics and science courses taken by a student, as recommended in Chapter 5. As noted there, a slightly more sophisticated strategy is to use the information in course titles; thus, one might expect a student to be exposed to more algebra content in a course called "Algebra" than in a course called "General Mathematics." A variety of analysts have related course-taking to individual achievement and found consistent and important relationships, independent of other measured student characteristics such as prior achievement and social class (see Raizen and Jones, 1985; Rock et al., 1985). These effects of course-taking swamp the effects of variables such as sex, race, public or private schooling, and teacher characteristics. Using the approach of logging the number of courses taken, the National Longitudinal Survey of 1972 and the High School and Beyond Survey of 1980 have provided data for very crude national or state indicators of content coverage in mathematics and science (National Center for Education Statistics, 1981, 1984). Because participation in high school courses is often up to students, the development of this indicator is discussed in Chapter 5.

In elementary school, the analogous measure to course-taking is a measure of the time devoted by the teacher to instruction in mathematics or science. This might be expressed on an absolute scale, such as number of minutes, or on a relative scale, such as percentage of the school day. Each would supply somewhat different information. A national or state indicator would require aggregating information across a representative sample of classrooms. Sometimes the nature of this information can be shocking: the most recently available survey of the time elementary school teachers (K–3) spend teaching science revealed that the average time per week was 17 minutes (Weiss, 1978). Also surprising is the variation from class to class in the amount of time spent on mathematics, generally considered a core subject in elementary school—from 23 to 61 minutes per day in two different fifth grades (Berliner, 1978). Because the amount of time allocated to mathematics or science instruction in a classroom is often the choice of the teacher, this measure is discussed in the chapters on both student behavior (Chapter 5) and teaching quality (Chapter 6). Once measures of the coverage of the actual curriculum are developed, measures of course-taking and time spent in instruction in mathematics and science might become superfluous. In the meantime, however, these measures are useful and relatively easy to gather.

A final, related measure discussed in Chapter 5 is the amount of time students spend on mathematics and science homework. A substantial body of literature finds that the careful use of homework enhances the learning of students; the amount of homework, the way that it is treated by the teacher, and its relationship to the curriculum all influence its effects (see Walberg, 1984; Raizen and Jones, 1985). Because homework is an expression of student behavior but also strongly affected by the teacher, this measure is first discussed in Chapter 5 on student behavior but further amplified in Chapter 6 on teaching quality.

DEVELOPING INDICATORS OF CONTENT COVERAGE

This section describes the approach that we recommend for developing indicators of content coverage. Succeeding sections consider indicators of curriculum quality including content depth, scientific accuracy, and pedagogical quality.

Curriculum Frameworks

We envisage the development of indicators for content coverage as starting with the development of exemplary curriculum frameworks. The frameworks would be intended to capture an "ideal" conception of the curriculum. They would be designed by a national group or groups; a variety of science organizations have the expertise and have produced curriculum recommendations over the years (Harms and Yager, 1981; American Chemical Society, 1984; Joint Committeee on Geographic Education, 1984). Also relevant may be state guidelines (California Department of Education, 1984; South Carolina Department of Education, 1986; Virginia Department of Education, 1986) and curricula from other countries (Klein and Rutherford, 1985; Travers et al., 1985). The mathematical community has been particularly active in thinking through the content of the school mathematics curriculum, as exemplified by the efforts of the Conference Board of the Mathematical Sciences (1983), the Mathematical Sciences Education Board (1987), and the National Council of Teachers of Mathematics (1987). California, Illinois, and Wisconsin also have constructed detailed frameworks for mathematics instruction in their school districts.

The objective envisaged by the committee would be to have a single national framework by grade cluster for each subject without dictating the placement of specific topics. Revised on a regular basis to reflect changes in the subject and advances in pedagogy, the frameworks would act as templates against which the content of existing and planned curricula could be matched. They would operate as a standard—that is, when a curriculum was mapped onto the framework, its content could be expressed in a measure representing the comprehensiveness of coverage of the curriculum. The measures might be expressed as a percentage of the content coverage represented by the framework. Taken alone, such a measure might be of some use to local and state policy makers. For example, suppose that local school officials wanted to buy a new K–5 textbook series and related materials in mathematics. They might use the measures of coverage of different textbook series in making their decision. Or, if a particular textbook series with limited content coverage were widely used in a certain state, the measure of content coverage might help explain to a state legislator why the children of the state scored badly on mathematics achievement tests compared with children in other states, although such an inference might or might not be correct.

Once measures of the content coverage of textbooks are developed for the major textbooks, for example, they could be weighted by the number of students using the textbooks to develop aggregate measures of content coverage for a state or local district, or even for the nation. Such aggregate measures might help a state legislator somewhat more in the quest to understand the low scores in the state. Moreover, if the content of the tests used in the states were also matched against the framework, then the content of the textbook series and the content of the tests would be expressed in the same nomenclature and a measure could be developed of the degree of overlap between the two. This might go even further in explaining the differences between the states in their test scores. As noted, a comparison of different eighth-grade curricula and their match with the content of the IEA tests explained much of the differences in results on the IEA tests (Crosswhite et al., 1985). Finally, a state legislator might want to know how the state's guidelines compared with the framework.

Perhaps more important than measures of the content coverage of the intended curriculum would be measures of the coverage of the content of the actual curriculum taught to students. If such measures were also available, the state legislator could compare the degree of match between the content of the textbook curriculum, the test, and the actual curriculum. If the content of the textbook and the test matched and they both fit with the state guidelines but the actual curriculum did not match, perhaps the suggested policy would be to increase teacher in-service education rather than to change the textbooks or tests.

And, as another example, if the content of the state guidelines were mapped against the framework, analyses could be carried out of how faithful the content of the textbooks and the actual curriculum were to the wishes of the state policy makers who developed the guidelines.

Establishing Subject-Matter Frameworks

The usefulness of comparisons of curriculum materials with exemplary frameworks is directly related to the quality of these frameworks. We recommend that the effort to develop such frameworks be started immediately, be well funded, and not be hurried. If developed in the way that we propose, they would serve as touchstones (or ideal descriptions) for the development of future guidelines, textbooks, and tests. They deserve the best effort that the nation's scientific and educational community can give, including a continual process of review and revision.

In priority order, the committee recommends first establishing frameworks for science in grades K–5 and 6–8. These two areas of schooling require immediate attention. It is suggested that this be followed with establishing frameworks for mathematics in grades K–5 and 6–8 (or K–4 and 5–8), then with frameworks for science literacy and mathematics literacy in grades 9–12, and finally with the frameworks for college-bound youth in mathematics and science, grades 9–12. An important caveat needs to accompany these suggestions:

optimally, instruction in science and mathematics ought not be separated in such a rigid manner, quite the contrary. Particularly in the lower grades, an integrated curriculum would be highly desirable. Constructions of such curricula, however, need to build on efforts to define in some detail the substantive core of subject matter from each discipline that is appropiately taught—preferably in a related, if not integrated way—at the given grade levels.

The frameworks must represent the structures of the subject matter and desirable learning goals, or alternatives among desirable goals. The frameworks should meet some general criteria: they should array, in a two-dimensional or more complex format, major processes, emphases, or principles in the curriculum against content topics, rather than simply list detailed topics; they should represent the best thinking of a combination of disciplinary specialists and specialists in the design of curricula and in teaching the subject; they should be conceived to "lead" practice, rather than representing a least common denominator of current practice; and they should be flexible, presenting a commonly agreed-on core and allowing for major options or alternatives in the content presented in states, localities, schools, and classrooms.

The core should represent a detailed explication of the standards for scientific and mathematical literacy set out in Chapter 2, adjusted for the appropriate grade levels. Over time, these frameworks should be regularly and critically revisited, so they reflect developments in the discipline, in pedagogy, and in the nation's aspirations for its youth. It seems reasonable to expect that a major review of each of the frameworks be made every decade.

As noted, work on the frameworks could build on existing efforts. In mathematics, for example, a framework based on the taxonomy developed by Romberg (1983) might have three dimensions: one concerned with the activities common to all mathematics, one with the specific processes entailed in doing mathematics, and one with the conceptual strands of mathematics that represent the historical development and core of the field. Romberg lists the essential activities common to all mathematics as:

- abstracting (i.e., dealing with quantitative relations and spatial forms to the exclusion of all other properties of objects);
- inventing (e.g., dealing with complex tasks with nonobvious solutions, making guesses or assertions and then demonstrating their logical validity);

- proving (using fundamental concepts to deduce a theorem through logical argument); and
- applying (using mathematics in the sciences, in engineering, in business and industry, in private and social life).

Four basic sets of processes involved in doing mathematics need to be included in the curriculum:
- relation processes—describing, classifying, comparing, ordering, separating, grouping, and partitioning;
- representation processes—going from the concrete to the abstract (or from the abstract to the concrete) in solving problems;
- symbolic-procedure processes—for example, the common algorithms learned in elementary school; and
- validation processes—carried out through empirical or logical deductive determinations.

The seven strands of substantive content suggested by Romberg for the core curriculum are:
- whole numbers arithmetic—counting, addition and subtraction, and multiplication and division;
- spatial relations—basic concepts of geometry;
- measurement—relating numbers to geometry;
- fractions—extending the concept of number from whole numbers to fractional numbers;
- coordinate geometry—procedures for assigning numbers to points in any space;
- algebra—dealing with abstractions from concrete numbers; and
- statistics—bridging the world of mathematics and the world of practical problems through data analysis and the interpretation of various forms of data collection and display.

Often added to these strands are two more topics (see, for example, Conference Board of the Mathematical Sciences, [1983]):
- discrete mathematics—basic combinatorics, graph theory, discrete probability and
- computer science—programming, introduction to algorithms, iteration.

The advent of the personal computer and related technology makes it particularly critical to rethink the mathematics curriculum at this time. It calls for the introduction of such new subject matter as computer science; it makes possible new ways of teaching traditional mathematics; and it forces reconsideration of the place of many traditional topics and instructional strategies. The computer has materially changed the ways in which mathematics is being done. Therefore, its use is integral for teaching the kinds of curricula being suggested by mathematicians and mathematics educators.

The development of K–5 science curricula is far behind efforts in mathematics. The committee is particularly concerned about this area, since the foundation for scientific literacy must be laid in those years. Several states are now taking steps to remedy the virtual absence of science teaching from elementary school. An example is the Oregon Framework for Science Programs (Northwest Evaluation Association, 1986). The main components of this framework are:

- Scientific concepts (e.g., cause-effect, change, cycle)
- Scientific problem-solving and inquiry processes (e.g., classifying, hypothesizing, inferring)
- Applications of hand skills (e.g., measuring, constructing)
- Interests in science (e.g., scientific avocation, confidence)
- Values that underlie science (e.g., questioning, searching for data, considering consequences)
- Interactions between science and society (e.g., science's influence on society, limitations of science)
- Characteristics of science (e.g., tentative, replicable, probabilistic)

These components encompass the latter three dimensions of scientific literacy (see Chapter 2). An adequate framework would also need to include a common core and optional components of factual knowledge, concepts, and theories representative of the first dimension of the literacy model, called in Chapter 2 "the nature of the scientific world view." (See, for example, the longer-term effort of the American Association for the Advancement of Science [1985b] to examine what science and technology is most worth learning.)

Obtaining Measures of Content Coverage

Once the exemplary frameworks are established, the next task is to map the content of the various exemplifications of curriculum onto the framework to derive measures of content coverage. For the most part, this rating task would have to be carried out jointly by expert scientists and educational experts. As suggested in Chapter 3, a considerable amount of training should be given to the raters, and acceptable levels of coding reliability should be established. In particular, there may be legitimate reasons for giving different weights to the coverage of different topics and to the emphasis on facts, principles, and procedural knowledge. These reasons should be made quite explicit ahead of time; weights should be agreed to by any particular panel before rating begins or spelled out by individual raters; both the process of rating and the weights used in the application of a particular framework should be described in detail in reporting the panel's findings.

State Guidelines The task of mapping the 50 state guidelines onto any framework seems relatively straightforward. Indeed, in a number of states, it would be a trivial task since very sparse, or in some cases no, guidelines exist for some of the curriculum areas. One issue that would arise would be what document to take as the state guidelines if multiple sets occurred in legislative, regulatory, and subregulatory material. The answer to this would have to be worked out on a state-by-state basis. It would also be useful to map the national guidelines of countries such as Japan, West Germany, and France, and the guidelines of the provinces of Canada for later purposes of comparison.

Textbook Series The task of mapping the content of textbook series and their related materials (laboratory exercises, computer software, films, workbooks) would be more tedious but straightforward. One decision here would have to do with which textbooks to assess. Two types seem important for national purposes: textbooks that are widely used and textbooks that are reputed to be exemplary themselves. States and local districts may wish to select textbooks and ancillary materials for mapping that are being considered for adoption or local purchase.

Tests Similarly, mapping test content onto a framework and deriving measures of coverage appear routine. Again, we suggest that frequency of use and reputation determine the tests to be chosen for analysis at the national level. Tests being considered for use at the state or local level should be chosen for analysis at these levels.

Implemented Curriculum The methodology for assessing the curriculum as actually implemented is more problematic. One way to do this would be to observe classes, an expensive and time-consuming method. Some members of the committee believe that teachers could supply adequate information. Presented with a list of items in the subject, a sample of teachers would be asked whether they covered the topic; items in the list would be drawn purposefully from the content framework but would be presented to teachers as an unorganized list. Teachers would be asked if they covered the topic that year, whether it was covered previously, or whether it is not covered at all in their school. This is similar to the approach used to assess the "opportunity to learn" measures in the IEA studies (Crosswhite et al., 1985; Jacobson, 1985) and by several investigators studying classroom processes (Barr, 1985; McLean, 1985). (For a description of this methodology, see Raizen, 1987.) A simple matrix sampling design would make it possible for each teacher to respond to only a few of the questions regarding the coverage of subject matter in his or her classroom and yet make possible estimates of coverage for the total framework.

Other members of the committee agreed that observation would be expensive but believed that teachers might not respond accurately or might forget what they had or had not taught. Research needs to be carried out to establish the validity of the approach of using teacher-reported information by conducting cross-checks with classroom observation conducted by outside observers. Research based on classroom observation could also probe the current ambiguity surrounding the meaning of "covering" a topic or subject.

Frequency of Mapping The mapping of state guidelines, textbook series, and test areas would require periodic updating as the content is changed or new materials are developed. The sampling of the actual content of instruction should be carried out nationally at least every four years on a cycle that is synchronized with the cycles for student assessment, so that the resulting indicators could be used together.

Recommendations

Research and Development: In order to develop indicators of breadth of content coverage in the science and mathematics curriculum, the committee recommends that exemplary frameworks be constructed for the following curriculum blocks: grades K–5 science, grades K–5 mathematics, grades 6–8 science, grades 6–8 mathematics, grades 9–12 literacy in science, grades 9–12 literacy in mathematics, grades 9–12 science for college-bound students, and grades 9–12 mathematics for college-bound students. The frameworks for grades K–5 and 6–8 science should be accorded the highest priority.

The frameworks must represent the structures of the subject matter and desirable learning goals, or alternatives among desirable goals.

Key Indicator: Once the frameworks are constructed, the committee recommends that three elements of the intended curriculum should be matched and rated against them for content coverage: state guidelines, textbooks and such associated materials as computer software and laboratory exercises, and tests. The frameworks should also be used to analyze the content coverage of the implemented curriculum (i.e., the content presented to the student as reported by classroom teachers).

The ratings obtained through analysis of the three elements of the intended curriculum and analysis of the implemented curriculum will provide the raw material for the construction of indicators of content coverage. The ratings should be carried out every four years at the national level in synchronization with the student assessments recommended in Chapter 4 so that the indicators can be used together. Ratings could be aggregated in different ways for different uses and different policy makers.

Research and Development: The committee recommends that research be carried out to establish the validity of teacher-reported information regarding content coverage in the classroom.

DEVELOPING INDICATORS OF CURRICULUM QUALITY

All three dimensions of curriculum quality—depth of topic treatment, scientific accuracy, and pedagogic quality—present difficult problems of measurement since they do not lend themselves to the kind of detailed analysis suggested for assessing breadth of coverage. The judgment of experts is required. For each of these quality dimensions, we discuss below why and how it should be assessed, together with some implications for developing assessment criteria.

Depth of Treatment

Discussions of science education have emphasized that a major goal is to give students a basic understanding and appreciation of the structure of, say, physics as a scientific discipline, the process of doing physics, and some of the complex problems solved and created by its applications. The ability to pass tests emphasizing the recall of science facts is not a sufficient foundation either for general scientific literacy or for further study. Therefore, the curriculum needs to concentrate on a limited number of topics to be studied in depth, in contrast to the makeup of many textbooks that, with every revision, add more topics to an already overburdened curriculum (Hurd et al., 1981; Taylor, 1984). The topics ought to be carefully chosen so that their presentation forms a coherent body of knowledge. Not only will this enhance learning with understanding; research suggests that in-depth study of particular topics as well as the use of laboratory or hands-on experiences are related to the engagement of students and to their interest in a course (Harms and Yager, 1981). Arguably, positive experiences in science classes engendered by these instructional strategies influence later interests and involvement in science.

The burden of this argument is that frameworks need to accommodate judgments on depth of coverage as well as breadth of coverage. How might this be accomplished? The most important requirement is that depth of coverage be an explicit evaluation criterion. Then, once a framework is in hand and the tasks of mapping of content coverage are complete, sets of judgments on the depth of coverage of text and other materials (or of reported or observed classroom practices) can be made depending on the weights assigned by the judges to the importance of various topics, concepts, and processes.

One sort of measure that might be used to assess depth of treatment is the number of pages devoted to a topic in a textbook, number of items on a test, or the amount of instructional time suggested in state guidelines or reported by teachers. Admittedly, these may turn out to be superficial measures; they certainly would need validation. It may well be that different judges might differ on the weights to be assigned to the treatment of particular topics and even to the need for broad coverage as contrasted to the depth of coverage of key topics and concepts. Scientists and science educators may place greater emphasis on depth as contrasted to the extent of coverage than do state and local authorities charged with developing overall curriculum guidelines that need to be endorsed by practicing educators and politicians. This underscores the importance of making explicit the weights assigned to various topics, concepts, and process skills by different expert groups.

Ratings of depth of treatment should be constructed for all three elements of the intended curriculum—state guidelines, texts and associated materials, and tests—as well as for the content actually presented to the student. As with the analysis of extent of content coverage, the latter will require special surveys of teachers and students supported by classroom observation.

Scientific Accuracy and Pedagogic Quality

The assessment of scientific accuracy of the content of the intended curriculum (state and local guidelines, textbooks and associated computer software and laboratory materials, and tests) would appear to be relatively straightforward. Panels of scientists could be convened periodically to review the content of these materials to ensure scientific accuracy. Optimally, these judgments would be made in conjunction with the ratings of materials for content coverage and depth, so that information on all three factors regarding a particular textbook or test would become available simultaneously.

Assessing the scientific accuracy of the content of the implemented curriculum—what the students actually receive in class—is much more difficult, for the reasons already stated. Classroom observation may be an appropriate tool, but at best it could provide information for only a limited number of classrooms. Another approach may be to establish some sort of threshold: minimally, one would expect a teacher to have the subject-matter knowledge necessary to teach the content defined by the framework at a particular level or for a particular subject. Thus, performance on the teacher tests of subject-matter competence recommended in Chapter 6 might be considered a minimum by which to judge the scientific accuracy of instruction, particularly if these tests were to be based on the content of the relevant framework. If tests of student learning were also based on the framework, it would be reasonable to expect teachers to perform well on the very same tests as a necessary, if not sufficient, indicator of subject-matter competence.

Similar problems arise in assessing pedagogic quality as in assessing scientific accuracy, even though the aspects of the curriculum to be judged are different. As to the intended curriculum and its components, panels of relevant experts could judge their pedagogic strengths and weaknesses: the appropriateness of the instructional strategies, given the subject matter and the grade level; the design and sequencing of topics to be taught; consonance with what is known about learning various scientific or mathematical constructs, processes, and skills; specific approaches to learners with different backgrounds and interests; and the like. Indicators of the pedagogic quality of actual classroom practice, however, would be difficult and expensive to obtain. The limitations of teacher surveys and classroom observation already discussed apply to this area with even more cogency, since classroom practice is more difficult to distill and describe succinctly and teachers' professional competence is involved. Moreover, as pointed out in Chapter 6, there are additional obstacles in the relative lack of consensus on best pedagogic practice, despite years of research on teaching effectiveness (e.g., Darling-Hammond and Hudson, 1986). Despite a considerable body of work, little of it has focused on the teaching of specific subject matter, at specific grade levels, to student populations with specific competencies, even though it is probably the case that effective teaching strategies are closely linked to context. One would hesitate at this time even to suggest some sort of test of pedagogic competence. Perhaps the follow-up work to the reports by the Holmes Group Consortium (1984), which recommends a year of professional education after graduation with a liberal arts degree, and by the Carnegie Forum on Education and the Economy (1986), which will aim to create a national board for teaching standards and teacher certification, will help build the needed understanding and consensus on what pedagogic knowledge teachers need to have and be able to orchestrate in given settings.

Developing Criteria for Assessing Quality

We have suggested above that criteria for judging curriculum quality would be developed by the different panels of experts as they assess the depth with which topics are treated, the scientific accuracy of the content, and the pedagogic strategies used in presenting it. As experience with such judgments accumulates, criteria can be expected to become more finely honed.

An approach to establishing standards that would facilitate the work of the panels is the analysis of high-quality programs by scientists and educators with a view toward providing models of excellence. First, outstanding science and mathematics programs would be selected, somewhat in the fashion of the *Focus on Excellence* series (National Science Teachers Association, 1983–1984), but in a more systematic manner to cover adequately the several curriculum blocks from grades K–5 through high school. The programs would then be described in some detail, with particular attention to the three quality dimensions. In preparation for selecting candidate programs and developing the descriptions, professionals (scientists, science educators, teachers, cognitive researchers) would be surveyed for judgments on the characteristics of a high-quality curriculum. Through the suggested selection and analyses, the characteristics of acknowledged high-quality programs would be made explicit and perhaps synthesized to provide several models. As the models and descriptions of their quality characteristics became available, panels could use them as a basis for creating criteria in assessing depth of treatment, scientific accuracy, and pedagogic quality. The importance of this strategy is that it would encourage panels to base their judgments on leading curricula rather than on the average content of science and mathematics instruction.

Recommendations

Research and Development: Standards of excellence should be developed based on the best of curricula in current use.

High-quality programs encompassing the curriculum blocks suggested above should be selected, profiled, and analyzed to provide models of excellence in depth of content coverage, scientific accuracy, and pedagogic soundness of science and mathematics curricula.

Key Indicator: The quality of the curriculum should be assessed by expert panels along three dimensions: depth of content treatment, scientific accuracy, and pedagogic soundness. Ratings for each of these quality dimensions should be assigned to the three elements of the intended curriculum (i.e., state guidelines, texts and associated materials, and tests). Assessments regarding depth of treatment should also be made of the implemented curriculum through teacher and student surveys and classroom observation.

To assess the depth of content treatment, the frameworks developed according to the recommendation made above should be used to identify the critical topics that constitute a coherent curriculum. Weights assigned by each rating panel regarding the depth of treatment desired for a given topic must be made explicit in reporting results.

The assessment of the scientific accuracy of the intended curriculum should be carried out by scientists in the relevant disciplines. The scientific content of the frameworks should be used to construct the tests of teacher competency of subject matter recommended in Chapter 6 and such tests used as a minimum measure of the scientific accuracy of the actual curriculum experienced by students.

Research and Development: The committee recommends research to provide validity checks on the standards being used to assess depth of treatment, scientific accuracy, and pedagogic soundness of science and mathematics curricula.

For example, research should be undertaken to establish what pedagogic knowledge teachers need to have and need to know how to use in order to teach science or mathematics effectively to students of different ages, backgrounds, and competencies.

IMPLICATIONS FOR STATE EDUCATION AGENCIES

Many state education agencies, for example, California (1984) and South Carolina (1986) in science and Texas in mathematics, are moving toward the curriculum framework concept described in this chapter. The Council of Chief State School Officers, representing all the state superintendents, has implicitly moved as an organization to a national framework by its endorsement of a state-by-state assessment system scheduled to be implemented in 1989 (Council of Chief State School Officers, 1984). Thus, a national framework could have an important function in making possible comparison and evaluations of the content of various state assessment tests in specific subjects.

The concept of a commonly agreed-on curriculum core allowing flexibility for alternatives reflects the commonality that really exists among schools, but preserves the cherished local and state freedom from federal curriculum control. The distinction between a "national" curriculum framework and a "federal" one is critical to states and localities—that is, the distinction between a set of guidelines developed by one or more nationally recognized groups and a prescribed course of studies mandated by a central authority. Because the proposed frameworks would be developed and applied for assessing curriculum content not just within a grade level or course, but over a reasonable period of schooling (e.g., the intermediate grades), there could be latitude regarding the sequencing of units. For example, a core topic might be taught in either sixth, seventh, or eighth grade. The framework concept would lead to a national grid of science and mathematics subject matter that would identify key concepts and processes to be included in the curriculum, but not the exact placement.

When such frameworks are developed, they can serve as a guide for a review of state level analyses including:

- Equal educational opportunity for all students regardless of socioeconomic status, location, race, ethnicity, or gender. Without accurate information on the breadth and depth of curriculum coverage as well as the variability among school systems, the first state indicator of a problem may be the number of small, rural school valedictorians who have to complete remedial mathematics courses before being accepted by the state's four-year colleges or the number of urban minority high school graduates who score at low levels in mathematics and science college placement tests.
- Guides for textbook analyses. Profiles matching the content analyses of textbooks with the appropriate framework could inform state and local decision makers about how well the textbooks will assist teachers in meeting the intended curriculum.
- Foundation for test development. The proposed curriculum frameworks could define the parameters for what a state believes all

students as well as the college-bound should know and be able to do as a result of their school experience in mathematics and science. Therefore, the frameworks could serve as the basis for a common understanding of what should be tested in a diagnostic way at the local school level and in a broader snapshot sense at the state level. At that point, the test results could serve as a basis for evaluating not only curriculum implementation but also whether the testing is sensitive enough to assess accurately the elements of the framework.

State and school district policies affect to some degree each classroom teacher's decisions on the implementation of a mathematics or science curriculum. The approach recommended by the committee provides a foundation for individuals at all levels to make informed decisions about what is working in the curriculum and about future directions.

8

Indicators of Financial and Leadership Support

RESOURCES AT THE LOCAL LEVEL

High-quality mathematics and science instruction requires significant financial support. As pointed out in Chapter 6, to attract and retain talented science and mathematics teachers, the schools must provide adequate salaries and the resources that teachers need to teach well. Moreover, good facilities, adequate time for planning instruction, and continuing professional development are necessary not only to attract talented teachers and support their teaching efforts, but also to support the development and use of curricula of high quality. Traditionally, financial support has been monitored through such indicators as expenditures per pupil, expenditures per student as a percentage of income per capita, average teacher salary, pupil/teacher ratio or pupil/staff ratio, federal funds as a percentage of school revenues, and the like (see, e.g., National Center for Education Statistics, 1985).

There are two reasons why the committee does not recommend the collection of additional information of this sort at the district level focused specifically on mathematics and science instruction. First, indicators of the particular resources purchased with school funds and how these resources are used to produce instruction, including time spent on specific subjects, provide more reliable evidence of the adequacy of financial support for mathematics and science instruction than do data on expenditures per pupil and other such traditional indicators of investments in education (Wiley and Harnischfeger, 1974; Denham and Lieberman, 1980; Levin, 1980). Second, the accounting systems in use in most American school districts do not permit the calculation of meaningful numbers on expenditures at the district level specifically supporting science and mathematics instruction. In a commissioned paper for the committee, Alexander (1985) described the requirements for a cost-accounting system that would be needed by local school districts to measure program costs for science and mathematics. According to Alexander (1985:16):

> An analysis of program costs for mathematics and science requires that expenditure components attributable to the respective programs be identified. Costs and expenditures are not synonymous. To find the actual costs of a particular program may involve expenditures from several budgetary components as well as indirect costs which must be prorated among programs or areas. . . . Costs for mathematics and science programs must necessarily derive from a school and course analysis. Costs for programs at the school district level would be derived from aggregation. True costs of programs can only be accurately determined by analysis at the school level.

The amount of time and effort that would be needed to develop, implement, and operate such a cost-analysis system in each local district is likely to discourage this approach to a financial indicator, particularly in view of the mixed findings in the literature on the connections between general educational expenditures and educational outcomes (see, e.g., Cohn and Riew, 1974).

After considerable discussion, the committee concluded that the best indicators would focus, not on dollars per se, but on the things money buys in a good educational program, namely, competitive salaries; the materials, supplies, and facilities needed to teach and learn well; time for teachers to plan instruction and engage in other professional activities; and opportunities provided to teachers for professional growth. Chapter 6 discusses these matters in greater detail and provides recommendations for the development of indicators of salaries, adequacy of working conditions, and availability and use of facilities, instructional materials, and supplies.

FEDERAL FINANCIAL SUPPORT

One way of gauging social commitment to an enterprise is to examine the amount of resources expended on it. Policy analysts (see, e.g., Wildavsky, 1979) have argued that the intent of public policy is to use public resources to achieve desired ends. The investment of federal funds in education certainly is a case in point. Concern with production of sufficient manpower in fields of perceived shortages has led to federal funding of fellowships for graduate and professional education; federal funds have supported the creation and maintenance of special educational programs for a variety of populations seen as underserved by the schools—poor children, children with physical handicaps and learning disabilities, children whose first language is not English, and children from minority ethnic groups (e.g., the recent program of magnet schools); in the 1960s, federal funds were invested in science and mathematics education to ensure a well-educated corps of scientists and engineers.

Although the federal financial contribution to elementary and secondary school science and mathematics instruction at any time has been small relative to state and local contributions, the federal government is in a unique position to exercise leadership—for example, by supporting the development of innovative curricula, by sponsoring educational and recognition programs for teachers, and by emphasizing that all children should have science and mathematics instruction of high quality. Therefore, indicators of federal support can provide important evidence of the social commitment to science and mathematics education (Catterall, 1986). The committee believes that it would be valuable to collect information annually on the level of federal financial support for elementary and secondary school science and mathematics instruction. This information should be broken down by discipline supported, school level (elementary, middle school, high school), and type of activity supported (materials development, teacher education/professional development, research and assessment, facilities and supplies, informal education, recognition programs, student activities).

Collecting information on the level of federal financial support of science and mathematics instruction is not straightforward. Some of the problems in obtaining reliable information on federal support were described by Mason (1985) in a paper written for the committee. Foremost is the fact that support comes from a number of federal agencies, but in several of the agency budgets, the dollars devoted to support for elementary and secondary school science and mathematics instruction do not appear as separate line items. Where data are available, they are found at two levels, at the macro-level of agency budgets and appropriations and at the micro-level of projects and activities.

An indicator of federal financial support based on macro-level data would include the portion of an agency budget specifically designated for mathematics and science education. For example, federal funds for education provided under the National Defense Education Act (NDEA) would have been included in a macro-level indicator, since the act included specific line item programs for improving precollege science and mathematics education, such as grants to public schools for laboratory and other special equipment used in teaching science and mathematics. A more recent example is the Education for Economic Security Act of 1983 (P.L. 98-377), which authorized financial assistance for state and local education agencies and institutions of higher education to improve the skills of teachers in mathematics, science, computer learning, and foreign languages.

The programs of the National Science Foundation (NSF) are the most visible and well-documented federal activities in science and mathematics education, and the NSF budget for education activities is often cited as an indicator of federal support. For the period from 1952 to 1980, science education obligations of NSF were reported according to function and level of education, with five main functional categories: research and development, students, teachers, institutions, and science and society. Currently, the precollege science education budget at NSF is organized as follows: materials development and informal education; teacher preparation and enhancement; and studies, research, and program assessment. The budget categories of NSF provide a useful starting point for developing indicators, but some analysis of project support would be necessary to report trend lines (Knapp et al., 1987).

It is not satisfactory to base an indicator of federal financial support on the NSF budget alone, since NSF is only one of several sources of federal support for science and mathematics education. However, an assessment of other federal financial support would require analyses of data at the micro-level of projects and activities administered by each agency. This kind of analysis can be conducted with varying degrees of ease or difficulty. For example, in the Department of Education, the largest grant programs to states and local districts are targeted broadly at students with special needs rather than at particular curricular areas. The Education for All Handicapped Children Act and Chapter 1 of the Education Consolidation and Improvement Act, which is targeted on disadvantaged children, provide funding for compensatory education. Although mathematics education is a major component of these programs, and, to a lesser degree, science and computer activities, no figures are currently available on dollars allocated to specific subjects. A special study would be needed to determine levels of funding by subject. Such a study would have to be sensitive to the problem of equating compensatory courses or activities in mathematics or science with regular instruction and making judgments on the extent to which compensatory education indicates improvement or decline in the quality of education in grades 1–12. Similarly, research and development supported by the Department of Education is not disaggregated by budget line items for specific subject areas. To develop an indicator, an analysis would be needed of the projects, grants, and contracts funded each year.

Other federal agencies, for example, the Department of Energy and the Department of Defense, may provide significant support for science and mathematics education in the schools and through out-of-school programs. Even with data at the micro-level of projects and activities, however, it would be difficult to make the needed distinctions within the budgets of these agencies for three reasons. First, funding data are not typically aggregated by function, and relevant projects or activities are not always identified as education projects or activities. Second, education activities, even when identified, are not necessarily classified as precollege or college-level activities. And third, science and mathematics education activities may not be distinguished from other subject areas. Thus, neither review of agency budgets nor analyzing lists of projects or activities may be effective in developing a reliable indicator of federal financial support for science and mathematics education in grades 1–12.

Categories of agency budgets tend to be highly generalized, and for most policy-analysis purposes it is necessary to obtain special cross-classifications or subcategories. The Office of Management and Budget (OMB) publishes a special analysis (K) on research and development from all agencies, but research and development related to science and mathematics education is not separately identified. Until a few years ago, OMB published a special analysis of education, but without breakdowns by subject area. According to OMB staff, there are no current plans for special analyses of education by subject (Bernard Martin, OMB, personal communication, August 1986).

Recommendation

Supplementary Indicator: The committee recommends the construction of a set of accounts detailing the level and type of support for science and mathematics education from all departments and agencies of the federal government that fund relevant programs.

The importance of having reliable annual data on the level of federal financial support merits the investment necessary to construct such a set of accounts. Agencies should be encouraged to report budget and funding data by categories identifiable as precollege mathematics and science education, and funds should be made available (possibly through NSF) to perform the necessary analyses. The kind of disaggregation of financial support for science and mathematics education found in the NSF budget could be used as a model for developing the recommended cross-agency indicator of federal support.

A somewhat similar argument could be made for a state-level indicator of financial investment in mathematics and science education. State policy makers continually have to make funding choices among all the curriculum areas. For example, should a program manager for state discretionary money direct the program staff for gifted students to emphasize the arts or science in its grant awards? Should policy makers influence program managers to use discretionary monies for staff development in reading or in mathematics? Should more mathematics and science specialists or consultants be hired? These state-level decisions not only demonstrate fiscal priorities but also send direct messages to local school personnel about what is important. Therefore, such decisions are also a way of describing the leadership role the state has taken in curriculum areas, particularly if there are discernible trends in financial support over time.

Financial support for student testing is another indicator of how important a curriculum area such as science is considered to be. For example, only half the states provide for state assessment of science knowledge, and the national assessment occurs at best every four years, whereas mathematics and reading are tested more frequently at both state and national levels.

While the committee has not suggested specific indicators of financial investment in mathematics and science education at the state or district level for the reasons indicated above, individual states and localities may wish to consider whether tracking such investments would give them useful information on curricular priorities.

NATIONAL LEADERSHIP

Support for elementary and secondary science and mathematics education at the national level should be measured not only in federal dollars but also in terms of the activities and efforts of the national scientific leadership. In the committee's view, the level of general social commitment to science and mathematics education needs to be motivated and shaped by the commitment of national leaders and leadership organizations of the scientific community (see Committee on Research in Mathematics, Science, and Technology Education, 1987). Examples from the past could be cited: the American Chemical Society, together with Glenn Seaborg, a national leader in science and education then and now, sparked the initiation and development of one of the major curriculum development projects of the 1960s (Seaborg, quoted in Committee on Research in Mathematics, Science, and Technology Education, 1987). The School Mathematics Study Group, probably the most influential curriculum-reform group of mathematics in the 1960s, was organized under the auspices of the American Mathematical Society, representing active researchers in mathematics, which "made it possible for a large number of distinguished college teachers and research mathematicians to enter wholeheartedly into cooperation with high school teachers in a concerted effort to improve the quality and presentation of school mathematics" (Wooton, 1965:13). It is not evident to what extent the scientific community remains involved in the improvement of science education. Since the success of any national effort will depend critically on the participation of scientific leaders, measures of the degree of their involvement are urgently needed.

While the interest and involvement of individual scientists in elementary and secondary education will always be idiosyncratic, the involvement of national scientific bodies ought to be constant and sustained. To monitor such commitment, a possible indicator might be the fraction of the staff and budget of relevant organizations that is devoted to advancing and improving elementary and secondary school science and mathematics education. These organizations include the American Association for the Advancement of Science, the National Academy of Sciences, the Mathematical Association of America, the American Institute of Physics, the American

TABLE 8-1 Investment in Education by the American Chemical Society

Level	Percentage	
	1986	1987
Elementary	7	8
High school	17	18
College	33	33
College and high school*	15	15
Other	28	26
Total funding	$985,000	$1,033,000

*Some programs serve both the college and high school communities.

Chemical Society, the American Institute of Biological Sciences, and the American Geological Institute. Some of these organizations have education divisions and ongoing projects of support to education; it would therefore be relatively easy to track increases and decreases in support over time. For example, the American Chemical Society intends to spend about 22.4 percent of its dues on education, divided as shown in Table 8-1.

The society has about 16 staff members who provide educational services supported by dues. In addition, in 1986, grant-supported programs provided for educational activities funded at $658,000; in 1987, this figure is expected to exceed $500,000. The society also operates self-sustaining activities budgeted for revenues of $2,249,000 in 1986 and $2,375,000 for 1987. This includes development and distribution of all kinds of educational materials such as newsletters, classroom curricular materials, comic books for elementary school children, textbook series for prospective chemistry technicians, and a variety of training programs (Kenneth Chapman, personal communication, September 26, 1986).

This sort of information should be available in a systematic way, but it is not. There are two reasons why it is important to obtain it on a continuing basis for individual fields of science and their associated professional bodies. First, if data on investments in education by scientific bodies were available periodically, one could track the level of involvement of the scientific community in the improvement of science education over time. Second, the efforts of individual professional societies could be compared with the needs in each field and with the efforts of their sister societies. In that connection, it also would be of interest to obtain an estimate of the pro-rated time of top executives and elected officials that is devoted to education-related activities. Such measures, after appropriate analysis, would provide evidence of changes in the extent to which the national scientific leadership devotes time and energy to improving science and mathematics instruction in elementary and secondary school.

Recommendation

Supplementary Indicator: The committee recommends that indicators be designed using budgetary data of scientific bodies and information on staff time and volunteer time devoted to education and that these indicators be routinely available to reflect the commitment of resources by scientific bodies for the improvement of mathematics and science education in the schools.

References

Aiken, L. R., Jr.
    1970   Attitude towards mathematics. *Review of Educational Research* 40:551–596.

Alexander, Kern
    1985   Costs of Precollege Mathematics and Science Programs: Analysis and Potential Use for Resource Allocation. Paper prepared for the Committee on Indicators of Precollege Education in Science and Mathematics, National Research Council.

American Association for the Advancement of Science
    1985a   Biology textbooks. *Science Books & Films* 20(5):245–286.
    1985b   Project 2061: Education for a Changing Future: Phase I. What understanding of science and technology will be important for everyone in tomorrow's world? Washington, D.C.: American Association for the Advancement of Science.
    1986a   Chemistry textbooks. *Science Books & Films* 21(5):249–293.
    1986b   Physics textbooks special. *Science Books & Films* 22(1):1–28.

American Chemical Society
    1984   Tomorrow: The Report of the Task Force for the Study of Chemistry Education in the United States. Washington, D.C.: American Chemical Society.

American Institute of Physics (AIP)
    1983   Salaries. AIP pub. no. R-311.01. Manpower Statistics Division. New York: American Institute of Physics.

REFERENCES

Anderson, C. W.
1985   Science Testing Programs and the Improvement of Teaching. Paper presented at the annual meeting of the American Educational Research Association, Chicago, Ill.

Anderson, C. W., and Smith, E. L.
1983   Children's Conception of Light and Color: Developing the Concept of Unseen Rays. Paper presented at the annual meeting of the American Educational Research Association, Montreal.

Anderson, J. R., Boyle, C. F., and Reiser, B. J.
1985   Intelligent tutoring systems. *Science* 228:450–462.

Angoff, W. H.
1984   Scales, Norms, and Equivalent Scores. Princeton, N.J.: Educational Testing Service.

Armor, David, et al.
1976   *Analysis of the School Preferred Reading Program in Selected Los Angeles Minority Schools.* Santa Monica, Calif.: Rand Corporation.

Assessment of Performance Unit
1974–   *Science Report for Teachers*: 1–5. Department of Education and
1975   Science, Welsh Office; Department of Education for Northern Ireland. Distributed by Garden City Press Limited, Letchworth, Hertfordshire SG6 1JS, Great Britain.

Ausubel, D. B.
1968   *Education Psychology: A Cognitive View.* New York: Holt, Rinehart & Winston.

Barr, Rebecca
1985   A Sociological Analysis of the Influence of Class Conditions on Mathematics Instructions. Paper presented at the annual meeting of the American Educational Research Association, Chicago. Available from the National College of Education, Evanston, Ill.

Barrows, H. S., Norman, G. R., Neufeld, V. R., and Feightner, J. W.
1982   The clinical reasoning of randomly selected physicians in general medical practice. *Clinical and Investigative Medicine* 5:49–55.

Becker, Henry Jay
1986   Instructional uses of school computers. *Reports from the 1985 National Survey* 1:1–12. Available from Center for Social Organization of Schools, Johns Hopkins University.

Begle, Edward G.
1979   *Critical Variables in Mathematics Education.* Washington, D.C.: Mathematical Association of America and National Council of Teachers of Mathematics.

Bell, A. W., Costello, J., and Kuchemann, D.
1983   *A Review of Research in Mathematical Education.* Windsor, Berks., United Kingdom: NFER-Nelson.

Berliner, David C.
1978   Allocated Time, Engaged Time, and Academic Learning Time in Elementary Mathematics Instruction. Paper presented at the meeting of the National Council of Teachers of Mathematics, San Diego, Calif.

Berman, Paul, and McLaughlin, Milbrey W.
1974– *Federal Programs Supporting Educational Change.* 8 Vols. Prepared
1975 for U.S. Department of Health, Education, and Welfare. Santa Monica, Calif.: Rand Corporation.

Bloom, Benjamin S.
1976 *Human Characteristics and School Learning.* New York: McGraw-Hill.
1984 The 2 sigma problem: The search for methods of group instruction as effective as one-to-one tutoring. *Educational Research* 13:4–16.

Blumberg, Fran, Epstein, Marion, MacDonald, Walter, and Mullis, Ina
1986 A Pilot Study of Higher-Order Thinking Skills Assessment Techniques in Science and Mathematics: Final Report. Parts 1 and 2. ME-G-84-2006-P4. Princeton, N.J.: National Assessment of Educational Progress.

Bransford, J., Sherwood, R., Vye, N., and Rieser, J.
1986 Teaching thinking and problem solving: Research foundations. *American Psychologist* 41:1078–1089.

Brophy, Jere E.
1986 Teacher influences on student achievement. *American Psychologist* 41:1069–1077.

Brophy, Jere E., and Good, Thomas L.
1986 Teacher behavior and student achievement. In *Third Handbook of Research on Teaching,* edited by M. C. Wittrock, 328–375. New York: Macmillan.

Brown, J. S., and Burton, R. R.
1978 Diagnostic models for procedural bugs in basic mathematical skills. *Cognitive Science* 2:123–146.

Brown, J. S., and VanLehn, K.
1980 Repair theory: A generative theory of bugs in procedural skills. *Cognitive Science* 4:379–426.

Brownell, William A.
1947 The place of meaning in the teaching of arithmetic. *Elementary School Journal* 47(January):256–265.

Bruner, Jerome S.
1960 *The Process of Education.* Cambridge, Mass.: Harvard University Press.
1966 *Towards a Theory of Instruction.* Cambridge, Mass.: Harvard University Press.

Bryk, Anthony S., Holland, Peter B., Lee, Valerie E., and Carriedo, Ruben A.
1984 *Effective Catholic Schools: An Exploration.* Washington, D.C.: National Catholic Education Association.

Burkhardt, Hugh
1986 *The Language of Functions and Graphs.* London: Joint Matriculation Board—Shell Centre for Mathematical Education, University of Nottingham.

Burton, Richard R., and Brown, John S.
1979 An investigation of computer coaching for informal learning activities. *International Journal of Man-Machine Studies* 11:5–24.

Byrne, C. J.
1983   Teacher Knowledge and Teacher Effectiveness: A Literature Review, Theoretical Analysis and Discussion of Research Strategy. Paper presented at the meeting of the Northwestern Educational Research Association, Ellenville, N.Y.

Cain, G. C., and Watts, H. W.
1970   Problems in making inferences from the Coleman Report. *American Sociological Review* 31:228–242.

California Department of Education
1984   *Science Framework Addendum.* Sacramento: California Department of Education.

Carey, Neil
1986   Instructional Indicators in an Educational Monitoring System: Alternatives and Recommendations. Working draft prepared for the National Science Foundation. WD-2919/4-NSF. Rand Corporation, Santa Monica, Calif.

Carley, M.
1981   *Social Measurement and Social Indicators.* London: George Allen and Unwin.

Carnegie Forum on Education and the Economy
1986   *A Nation Prepared: Teachers for the 21st Century.* Report of the Task Force on Teaching as a Profession. New York: Carnegie Forum on Education and the Economy.

Catterall, James S.
1986   Resource Indicators and Social Commitment to Education. Working draft prepared for the National Science Foundation. WD-2919/8-NSF. Rand Corporation, Santa Monica, Calif.

Cavana, G. R., and Leonard, W. H.
1985   Extending discretion in high school science curricula. *Science Education* 69(5):593–603.

Cavin, Edward S.
1986   A Review of Teacher Supply and Demand Projection by the U.S. Department of Education, Illinois, and New York. Paper prepared for the Panel on Statistics on Supply and Demand for Precollege Science and Mathematics Teachers, Committee on National Statistics, National Research Council. Available from Mathematica Policy Research, Inc., Princeton, N.J.

Charters, W. W., Jr.
1970   Some factors affecting teacher survival in school districts. *American Educational Research Journal* 7(1):1–27.

Chi, M. T. H., Feltovich, P. J., and Glaser, R.
1981   Categorization and representation of physics problems by experts and novices. *Cognitive Science* 5:121–152.

Chipman, S. F., Brush, L. R., and Wilson, D. M., eds.
1985   *Women and Mathematics: Balancing the Equation.* Hillsdale, N.J.: Erlbaum.

Clancey, W. J.
1979   Tutoring rules for guiding a case method dialogue. *International Journal of Man-Machine Studies* 11:25–49.

Cohn, E., and Riew, J.
1974　　Cost functions in public schools. *Journal of Human Resources* (3):422–434.

Coleman, J. S., Campbell, E. Q., Hobsen, C. J., McPartland, J., Mood, A., Weinfield, F. D., and York, R. T.
1966　　*Equality of Educational Opportunity.* Washington, D.C.: U.S. Department of Health, Education, and Welfare.

Committee on Research in Mathematics, Science, and Technology Education
1985　　*Mathematics, Science, and Technology Education: A Research Agenda.* Available from the Commission on Behavioral and Social Sciences and Education. Washington, D.C.: National Academy Press.
1987　　*Interdisciplinary Research in Mathematics, Science, and Technology Education.* Available from the Commission on Behavioral and Social Sciences and Education. Washington, D.C.: National Academy Press.

Conference Board of the Mathematical Sciences
1983　　The mathematical sciences curriculum K–12: What is still fundamental and what is not. In *Educating Americans for the 21st Century: Source Materials.* National Science Board Commission on Precollege Education in Mathematics, Science, and Technology. Washington, D.C.: National Science Foundation.

Confrey, Jere
1985　　A Constructivist View of Mathematics Instruction: A Theoretical Perspective. Paper presented at the annual meeting of the American Educational Research Association, Chicago, Ill.

Connecticut State Department of Education
1985　　*Teacher Supply and Demand in Connecticut.* April 4, 1985. Hartford: Connecticut State Department of Education.

Connell, J. P.
1985　　A new multidimensional measure of children's perceptions of control. *Child Development* 56:1018–1041.

Connell, J. P., and Ryan, R. M.
1984　　A development theory of motivation in the classroom. *Teacher Education Quarterly* 11(4):64–77.

Cooley, W., Bond, L., and Mao, B. J.
1981　　Analyzing multi-level data. In *Educational Evaluation Methodology*, edited by R. Berk. Baltimore: Johns Hopkins University Press.

Council of Chief State School Officers
1984　　*Education Evaluation and Assessment in the United States.* Position Paper and Recommendations for Action. Washington, D.C.: Council of Chief State School Officers.

Cronbach, L. J., and Meehl, P. E.
1955　　Construct validity in psychological tests. *Psychological Bulletin* 52:281–302.

Cronbach, L. J., and Snow, R. E.
1977　　*Aptitudes and Instructional Methods: A Handbook for Research on Interactions.* New York: Irvington.

Crosswhite, F. Joe
1972　　*Correlates of Attitudes Toward Mathematics.* National Longitudinal Study of Mathematical Abilities Report No. 20. Pasadena, Calif.: A. C. Vroman.

Crosswhite, F. Joe, Dossey, John A., Swafford, Jane O., McKnight, Curtis C., and Cooney, Thomas J.
    1985   *Second International Mathematics Study Summary Report for the United States.* Champaign, Ill.: Stipes.

Darling-Hammond, Linda
    1984   *Beyond the Commission Reports: The Coming Crisis in Teaching.* Report No. R-3177-RC. Santa Monica, Calif.: Rand Corporation.

Darling-Hammond, Linda, and Hudson, Lisa
    1986   Indicators of Teacher and Teaching Quality. Working Draft prepared for the National Science Foundation. WD-2919/5-NSF. Rand Corporation, Santa Monica, Calif.

Darling-Hammond, Linda, Haggstrom, Gus, Hudson, Lisa, and Oakes, Jeannie
    1986   A Conceptual Framework for Examining Staffing and Schooling. Draft report prepared for the Center for Education Statistics. Rand Corporation, Santa Monica, Calif.

deGroot, A. D.
    1965   *Thought and Choice in Chess.* The Hague: Mouton.

Denham, Carolyn, and Lieberman, Ann, eds.
    1980   *Time to Learn.* Available from the U.S. Government Printing Office. Washington, D.C.: National Institute of Education.

Diekoff, G. M.
    1983   Testing through relationship judgments. *Journal of Educational Psychology* 75:227–233.

Dienes, Z. P., and Golding, E. B.
    1971   *Approach to Modern Mathematics.* New York: Herder and Herder.

Druva, Cynthia A., and Anderson, Ronald D.
    1983   Science teacher characteristics by teacher behavior and by student outcome: A meta-analysis of research. *Journal of Research in Science Teaching* 20(5):467–479.

Eberts, Randall W., and Stone, Joe A.
    1984   *Unions and Public Schools.* Lexington, Mass.: Lexington Books.

Egan, D., and Schwartz, B.
    1979   Chunking in recall of symbolic drawings. *Memory and Cognition* 7:149–158.

Embretson (Whitley), S.
    1983   Construct validity: Construct representation versus nomothetic span. *Psychological Bulletin* 93:179–197.

Ericsson, K. A., and Simon, H. A.
    1984   *Protocol Analysis: Verbal Reports as Data.* Cambridge, Mass.: MIT Press.

Evertson, C., Anderson, D., Anderson, L., and Brophy, Jere
    1980   Relationships between classroom behaviors and student outcomes in junior high mathematics and English classes. *American Educational Research Journal* 17:43–60.

Flanagan, John C., and Cooley, William W.
    1966   *Project Talent One-Year Follow-Up Studies.* Pittsburgh: School of Education, University of Pittsburgh.

Flowers, A.
    1984   Preparation of teachers: Myths and realities. In *Teacher Shortage in Science and Mathematics: Myths, Realities and Research*, edited by J. L. Taylor. Washington, D.C.: National Institute of Education.

Fraser, Barry S., Welch, Wayne W., and Walberg, Herbert J.
1986   Using secondary analysis of national assessment data to identify predictors of junior high school students' outcomes. *The Alberta Journal of Educational Research* 32(1):37–50.

Frederiksen, C. H.
1975   Representing logical and semantic structure of knowledge acquired from discourse. *Cognitive Psychology* 7:371–485.
1984   Frame Construction in Children's Discourse Communication. Paper presented at the meeting of the American Psychological Association, Toronto, August.
1985   Cognitive models and discourse analysis. In *Written Communication Annual, Vol. 1: Linguistic Approaches to the Study of Written Discourse*, edited by C. R. Cooper and S. Greenbaum. Beverly Hills, Calif.: Sage Publications.

Frederiksen, C. H., Frederiksen, J. R., and Bracewell, R. H.
1985   Discourse analysis of children's text production. In *Writing in Real Time*, edited by A. Matsuhasi. New York: Longmans.

Frederiksen, J. R.
1982   A componential theory of reading skills and their interactions. In *Advances in the Psychology of Human Intelligence*, edited by R. J. Sternberg. Vol. 1. Hillsdale, N.J.: Erlbaum.

Frederiksen, N.
1984a   The real test bias: Influences of testing on teaching and learning. *American Psychologist* 39:193–202.
1984b   Implications of cognitive theory for instruction in problem solving. *Review of Educational Research* 54:363–407.
1986   Construct validity and construct similarity: Methods for use in test development and test validation. *Multivariate Behavioral Research* 21:3–28.

Frederiksen, N., and Ward, W. C.
1978   Measures for the study of creativity in scientific problem solving. *Applied Psychological Measurement* 2:1–24.

Freeman, R. B.
1976   *The Overeducated American*. New York: Academic Press.

Freudenthal, Hans
1983   Major problems of mathematics education. In *Proceedings of the Fourth International Congress of Mathematical Education*, edited by Marilyn Zweng, Thomas Green, Jeremy Kilpatrick, Henry Pollak, and Marilyn Suydams. Boston: Birkhäuser.

Gagné, R. M.
1965   *The Conditions of Learning*. New York: Holt, Rinehart & Winston.

Gardner, P. L.
1975   Attitude measurement, a critique of some recent research. *Education Research* 7:101–109.

Gentner, D., and Gentner, D. R.
1983   Flowing water or teeming crowds: Mental models of electricity. In *Mental Models*, edited by D. Gentner and A. L. Stevens. Hillsdale, N.J.: Erlbaum.

Glaser, Robert
1984   Education and thinking: The role of knowledge. *American Psychologist* 39(2):93–104.

Goertz, Margaret
 1986 *State Educational Standards: A 50-State Survey.* Princeton, N.J.: Educational Testing Service.
Goertz, Margaret E., Ekstrom, Ruth B., and Coley, Richard J.
 1984 *The Impact of State Policy on Entrance Into the Teaching Profession.* Princeton, N.J.: Educational Testing Service.
Goldstein, I.
 1980 Developing a computational representation for problem-solving skills. In *Problem Solving and Education: Issues in Teaching and Research,* edited by D. T. Tuma and F. Reif, 53–79. Hillsdale, N.J.: Erlbaum.
Goldstein, Paul
 1978 *Changing the American Schoolbook.* Lexington, Mass.: D.C. Heath.
Good, Thomas L.
 1983 Recent classroom research: Implications for teacher education. In *Essential Knowledge for Beginning Educators,* edited by D. C. Smith. Washington, D.C.: American Association of Colleges for Teacher Education.
Good, Thomas L., and Grouws, Douglas
 1979 The Missouri mathematics effectiveness project: An experimental study in fourth grade classrooms. *Journal of Educational Psychology* 71(3):355–362.
Good, Thomas L., and Weinstein, Thona S.
 1986 Schools make a difference: Evidence, criticisms, and new directions. *American Psychologist* 41(10):1090–1097.
Goodlad, John I.
 1984 *A Place Called School.* New York: McGraw-Hill.
Greenburg, David, and McCall, John
 1974 Teacher mobility and allocation. *Journal of Human Resources* 9(Fall)4:480–502.
Gulliksen, H.
 1950 *Theory of Mental Tests.* New York: John Wiley & Sons.
Haggstrom, Gus W., Darling-Hammond, Linda, and Grissmer, David W.
 1986 Assessing Teacher Supply and Demand. Draft report prepared for the Center of Education Statistics. Rand Corporation, Santa Monica, Calif.
Hanford, George H.
 1986 Some caveats on comparing SAT scores. *Education Week* October 8:20.
Hanushek, Eric A.
 1972 *Education and Race.* Lexington, Mass.: D.C. Heath.
Harms, Norris C., and Yager, Robert E., eds.
 1981 *What Research Says to the Science Teacher.* Washington, D.C.: National Science Teachers Association.
Harris, C. W.
 1963 *Problems in Measuring Change.* Madison: University of Wisconsin Press.
Hayes-Roth, B.
 1977 Evolution of structures and processes. *Psychological Review* 84:260–278.

Hein, George E.
In press
The assessment of science learning in materials-centered science education programs. *Science and Children.*

Hilton, Peter
1986
Current trends in mathematics and future trends in mathematics education. In *The Monitoring of School Mathematics: Background Papers*, edited by Thomas A. Romberg and Deborah M. Stewart. Vol. 1. Madison: Wisconsin Center for Education Research, University of Wisconsin.

Holmes Group Consortium
1984
New Standards for Quality in Teacher Education. Proposal to the Secretary's Discretionary Program/U.S. Department of Education, the Ford Foundation, and the Carnegie Corporation of New York. College of Education, Michigan State University.

Horn, Elizabeth A., and Walberg, Herbert J.
1984
Achievement and interest as functions of quantity and level of instruction. *Journal of Education Research* 77(4):227–232.

Hueftle, Stacey J., Rakow, Steven J., and Welch, Wayne W.
1983
Images of Science: A Summary of Results from the 1981-82 National Assessment in Science. Minneapolis: Minnesota Research and Evaluation Center.

Hurd, Paul DeH., Robinson, James T., McConnell, Mary C., and Ross, Norris M., Jr.
1981
The Status of Middle School and Junior High School Science. Biological Sciences Curriculum Study, Colorado Springs, Colo.

Husén, Torsten
1967
International Study of Achievement in Mathematics: A Comparison of Twelve Countries. Vols. I and II. New York: John Wiley & Sons.

Illinois State Board of Education
1983
The Supply and Demand for Illinois Mathematics and Science Teachers. Springfield: Illinois State Board of Education.

Institute for Research in Social Behavior
1984
University of California Faculty Time-Use Study. Report for the 1983–1984 Academic Year. Oakland, Calif.: Institute for Research in Social Behavior.

Jacobson, Willard J.
1985
National Report: U.S.A. Second International Science Study. New York: Teachers College, Columbia University.

Jaques, Elliott
1956
Measurement of Responsibility. London: Tavistock Publication. Reprinted in 1972 as *Measurement of Responsibility: A Study of Work, Payment, and Individual Capacity.* New York: John Wiley & Sons.

Joint Committee on Geographic Education
1984
Guidelines for Geographic Education. Washington, D.C.: Association of American Geographers.

Jones, Lyle V., Davenport, Ernest C., Jr., Bryson, Aloha, Bekhuis, Tanja, and Zwick, Rebecca
1986
Mathematics and science test scores as related to courses taken in high school and other factors. *Journal of Educational Measurement* 23(3):197–208.

Juster, F. Thomas, and Stafford, Frank P., eds.
1985   *Time, Goods, and Well-being.* Ann Arbor, Mich.: Survey Research Center, Institute for Social Science, University of Michigan.

Klein, Margarete S., and Rutherford, F. James
1985   *Science Education in Global Perspective: Lessons from Five Countries.* Boulder, Colo.: Westview Press.

Knapp, Michael S., et al.
1987   *Opportunities for Strategic Investment in K–12 Science Education: Options for the National Science Foundation.* Summary Report; Volume 1: Problems and Opportunities; Volume 2: Groundwork for Strategic Investment. SRI Project No. 1809. Menlo Park, Calif.: SRI International.

Kuhs, Therese, Schmidt, William, Porter, Andrew, Floden, Robert, Freeman, Donald, and Schwille, John
1979   *A Taxonomy for Classifying Elementary School Mathematics Content.* East Lansing, Mich.: The Institute for Research on Teaching, Michigan State University.

Kulm, Gerald
1980   Research on mathematics attitude. In *Research in Mathematics Education*, edited by Richard J. Shumway. Reston, Va.: National Council of Teachers of Mathematics.

Kyllonen, P. C.
1986   *Theory-Based Cognitive Assessment.* Brooks Air Force Base, Tex.: Air Force Systems Command.

Langbein, L. I., and Lichtman, A. J.
1978   *Ecological Inference.* Sage University Papers on Quantitative Applications in the Social Sciences, Series 07-010. Beverly Hills, Calif.: Sage Publications.

Larkin, J. H.
1979   Information processing models and science instruction. In *Cognitive Process Instruction*, edited by J. Lochhead and J. Clement. Philadelphia: Franklin Institute Press.

Larkin, J., McDermott, L., Simon, D. P., and Simon, H. A.
1980   Expert and novice performance in solving physics problems. *Science* 208:1335–1342.

Lefcourt, H. M., VonBaeyer, C. L., Ware, E. E., and Cox, D. J.
1979   The multi-dimensional-multiattributional causality scale: The development of a goal specific locus of control scale. *Canadian Journal of Behavioral Science* 11(4):286–304.

Levin, Henry M.
1980   Educational production theory and teacher inputs. In *The Analysis of Educational Productivity: Issues in Macroanalysis*, edited by C. E. Bidwell and D. M. Windham. Vol. 2. Cambridge, Mass.: Ballinger.

Lightfoot, Sara L.
1983   *The Good High School: Portraits of Character and Culture.* New York: Basic Books.

Lindquist, E. G.
1954   The Iowa electronic test processing equipment. In *1953 Invitational Conference on Testing Problems*, 160–168. Princeton, N.J.: Educational Testing Service.

Linn, Robert L.
1986 Educational testing and assessment: Research needs and policy issues. *American Psychologist* 41(10):1153–1160.

Lord, F. M.
1980 *Application of Item-Response Theory to Practical Testing Problems.* Hillsdale, N.J.: Erlbaum.

Lord, F. M., and Novick, M. R.
1968 *Statistical Theories of Mental Test Scores.* Reading, Mass.: Addison-Wesley.

Malone, T. W.
1981 Toward a theory of intrinsically motivating instruction. *Cognitive Science* 4:333–369.

Manski, Charles F.
1985 Academic Ability, Earnings, and the Decision to Become a Teacher: Evidence from the National Longitudinal Study of the High School Class of 1972. Working Paper No. 1539. National Bureau of Economic Research, Cambridge, Mass.

Mason, Ward S.
1985 Indicators of Federal Investment in Precollege Science, Mathematics, and Technology Education. Paper prepared for the Committee on Indicators of Precollege Science and Mathematics Education, National Research Council.

Mathematical Sciences Education Board
1987 Annual Report of the Mathematical Sciences Education Board. Available from the National Research Council, Washington, D.C.

McDermott, Lillian C.
1984 Research on conceptual understanding in mechanics. *Physics Today* July:24–32.

McKnight, Curtis C., Crosswhite, F. Joe, Dossey, John A., Kifer, Edward, Swafford, Jane O., Travers, Kenneth J., and Cooney, Thomas J.
1987 *The Underachieving Curriculum: Assessing U.S. School Mathematics from an International Perspective.* Champaign, Ill.: Stipes.

McLean, Les
1985 Drawing Implications of Instruction from Item, Topic and Classroom-Level Scores in Large-Scale Science Assessment. Paper presented at the annual meeting of American Educational Research Association, Chicago, Ill.

McLeod, D. B.
1986 Affective Influences on Mathematical Problem Solving. Paper prepared for the Conference on Affective Influences on Mathematical Problem Solving, San Diego, Calif., May 28–30.

Medrich, Elliot, Rorzin, Judith, Rubin, Victor, and Buckle, Stuart
1982 *The Serious Business of Growing Up: A Study of Children's Lives Outside of School.* Berkeley: University of California Press.

Messick, Samuel J.
1975 The standard problem: Meaning and values in measurement and evaluation. *American Psychologist* 30:955–966.
1980 *The Effectiveness of Coaching for the SAT: Review and Reanalysis of Research from the Fifties to the FTC.* Princeton, N.J.: Educational Testing Service.

Messick, S., Beaton, A., and Lord, F.
- 1983 *National Assessment of Educational Progress Reconsidered: A New Design for a New Era.* Princeton, N.J.: Educational Testing Service.

Meyer, D. E., and Schvaneveldt, R. W.
- 1976 Meaning, memory structure, and mental processes. *Science* 192:27–33.

Miller, Jon D.
- 1983 Scientific literacy: A conceptual and empirical review. *Daedalus* 112(2):29–48.
- 1986 Scientific Literacy in the United States. Paper presented at a symposium on Communicating Science to the Public, London. Available from Northern Illinois University.

Munby, Hugh
- 1983 Thirty studies involving the Scientific Attitude Inventory: What confidence can we have in this instrument? *Journal of Research in Science Teaching* 20(2):141–162.

Murnane, Richard J.
- 1975 *Impact of School Resources on the Learning of Inner City Children.* Cambridge, Mass.: Ballinger.
- 1981 Interpreting the evidence on school effectiveness. *Teachers College Record* 83(1):19–35.
- In press Understanding teacher attrition. *Harvard Educational Review.*

Murnane, Richard J., and Phillips, Barbara R.
- 1981 Learning by doing, vintage, and selection: Three pieces of the puzzle relating teaching experience and teaching performance. *Economics of Education Review* 4(Fall):453–465.

Naisbitt, J.
- 1982 *Megatrends: Ten New Directions Transorming Our Lives.* New York: Warner Books.

National Assessment of Educational Progress (NAEP)
- 1982 Graduates may lack tomorrow's "basics." *NAEP Newsletter* 15:8.
- 1987 *Learning by Doing.* A Manual for Teaching and Assessing Higher-Order Thinking in Science and Mathematics. Princeton, N.J.: National Assessment of Educational Progress.

National Center for Education Statistics
- 1981 *A Capsule Description of High School Students: A Report on HIgh School and Beyond, A National Longitudinal Study for the 1980s.* Prepared by Samuel S. Peng, William B. Fetters, and Andrew J. Kolstad. Supt. of Doc. No. 0-729-575/2100. Available from the U.S. Government Printing Office. Washington, D.C.: U.S. Department of Education.
- 1984 *High School Seniors: A Comparative Study of the Classes of 1972 and 1980.* Prepared by William B. Fetters, Jeffrey A. Owings, and George H. Brown. Supt. of Doc. No. 065-000-00204-3. Available from the U.S. Government Printing Office. Washington, D.C.: U.S. Department of Education.
- 1985 *The Condition of Education.* Prepared by Valena W. Plisko and Joyce D. Stern. Supt. of Doc. No. NCES 85-402. Available from

the U.S. Government Printing Office. Washington, D.C.: U.S. Department of Education.

National Council of Teachers of Mathematics (NCTM)
1987 Commission will develop NCTM standards for school mathematics. *News Bulletin* 23(3):3.

National Science Resources Center
1986 National Conference on the Teaching of Science in Elementary Schools. Summary. Available from the National Science Resources Center, National Academy of Sciences–Smithsonian Institution, Washington, D.C.

National Science Teachers Association
1983– *Focus on Excellence.* Vol. 1(1–5). John E. Penick et al., eds.
1984 Washington, D.C.: National Science Teachers Association.

New York State Education Department
1983 Teachers in New York State—1968 to 1982. State University of New York, State Education Department, Information Center on Education, Albany.

Newell, A., and Simon, H. A.
1972 *Human Problem Solving.* Englewood Cliffs, N.J.: Prentice-Hall.

Norman, G. R., Muzzin, L. J., Williams, R. G., and Swanson, D. B.
1985 Simulation in health sciences education. *Journal of Instructional Development* 8:11–17.

Northwest Evaluation Association
1986 A Handbook for Science Curriculum Development: Using the Science Curriculum Planning Matrix. Oregon Department of Education, Salem.

Northwestern Endicott Report
1985 *Thirty-ninth Annual Report.* Prepared by Victor R. Lindquist. Number and Average Monthly Salaries for New Graduates by Field and Degree, 1984 and 1985. Evanston, Ill.: Northwestern University.

Oakes, Jeannie
1986 *Educational Indicators: A Guide for Policymakers.* OPE-01. Santa Monica, Calif.: Rand Corporation.

Osborne, R. J., and Wittrock, M. C.
1983 Learning science: A generative process. *Science Education* 67:489–508.

Owen, D.
1985 *None of the Above: Behind the Myth of Scholastic Aptitude.* New York: Houghton Mifflin.

Panel on Statistics on Supply and Demand for Precollege Science and Mathematics Teachers
1987 *Toward Understanding Teacher Supply and Demand: Priorities for Research and Development.* Interim Report. Available from the Commission on Behavioral and Social Sciences and Education. Washington, D.C.: National Academy Press.

Patel, V., and Frederiksen, C. H.
1984 Cognitive processes in comprehension and knowledge acquisition by medical students and physicians. In *Tutorials in Problem-Based Learning,* edited by H. G. DeVolder and H. Schmidt. Assen, Netherlands: VanGorcum.

Penick, John E., ed.
1983 *Focus on Excellence.* Vol. 1, No. 2. Elementary Science. Washington, D.C.: National Science Teachers Association.
Penick, John E., and Yager, Robert E.
1983 The search for excellence in science education. *Phi Delta Kappan* 64(8):621–623.
Peterson, Penelope
1979 Direct instruction reconsidered. In *Research on Teaching*, edited by P. Peter and H. J. Walberg. Berkeley, Calif.: McCutcheon.
Piaget, Jean
1954 *The Construction of Reality in the Child.* New York: Basic Books.
Popham, W. H.
1983 Measurement as an instructional catalyst. In *New Directions for Testing and Measurement: Measurement, Technology, and Individuality in Education*, edited by R. B. Ekstrom, 19–30. San Francisco: Jossey-Bass.
Preece, P. F.
1976 Mapping cognitive structure: A comparison of methods. *Journal of Educational Psychology* 68:1–8.
Purkey, Stewart C., and Smith, Marshall S.
1983 Effective schools—a review. *Elementary School Journal* 83(4):426–452.
Raizen, Senta A.
1987 Assessing the quality of the science curriculum. In *Forum 86: The Science Curriculum.* Washington, D.C.: American Association for the Advancement of Science.
Raizen, Senta A., and Jones, Lyle V., eds.
1985 *Indicators of Precollege Education in Science and Mathematics. A Preliminary Review.* Committee on Indicators of Precollege Science and Mathematics Education, National Research Council. Washington, D.C.: National Academy Press.
Rasch, G.
1960 *Probabilistic Models for Some Intelligence and Attainment Tests.* Copenhagen: Nielson & Lydicke (for Denmark's Paedagogiske Institute).
Reggia, J. A., Perricone, B. T., Nan, D. S., and Peng, Y.
1985 Answer justification in diagnostic expert systems—Part I: Abductive inference and its justification. *IEEE Transactions on Biomedical Engineering* BME-32:263–272.
Reiser, B. J., Anderson, J. R., and Farrell, R. G.
1985 Dynamic Student Modeling in an Intelligent Tutor for LISP Programming. Proceedings from the 1985 International Joint Conference on Artificial Intelligence, Los Angeles, Calif.
Research Triangle Institute
1985 1985 National Survey: Science and Mathematics Education. Teacher Questionnaire. Research Triangle Institute, Research Triangle Park, N.C.
Resnick, D. P., and Resnick, L. B.
1985 Standards, curriculum, and performance: A historical and comparative perspective. *Educational Researcher* 14:5–20.

Resnick, Lauren B.
1983   Mathematics and science learning: A new conception. *Science* 220(4):477–478.
1987   *Education and Learning to Think.* Washington, D.C.: National Academy Press.

Robinson, W. S.
1950   Ecological correlations and the behavior of individuals. *American Sociological Review* 15:351–357.

Rock, Donald A., Ekstrom, Ruth B., Goertz, Margaret E., and Pollack, Judith M.
1985   Determinants of Achievement Gain in High School. Briefing paper for U.S. Department of Education and the National Center for Education Statistics. Educational Testing Service, Princeton, N.J.

Romberg, Thomas A.
1983   A common curriculum for mathematics. In *Individual Differences and the Common Curriculum: Eighty-Second Yearbook of the National Society for the Study of Education*, 121–159. Chicago: University of Chicago Press.
1986   Measures of Mathematical Achievement: Problems and Influences. Paper prepared for the National Conference on the Influence of Testing on Mathematics Education, University of California at Los Angeles. Available from the University of Wisconsin, Madison.

Romberg, Thomas A., and Carpenter, T. P.
1986   Research on teaching and learning mathematics: Two disciplines of scientific inquiry. In *Third Handbook of Research on Teaching*, edited by Merlin C. Wittrock. New York: Macmillan.

Rosenholtz, S. J.
1985   Effective schools: Interpreting the evidence. *American Journal of Education* 93:352–388.

Rowe, Mary Budd
1979   Externality and Children's Problem Solving Strategies. Paper prepared under NIMH grant no. R01MH 25229. Available from the University of Florida, Gainesville.
1983   Science education: A framework for decision makers. *Daedalus* 112(2):123–142.

Sarason, Seymour B.
1985   The school as a social situation. *Annual Review of Psychology* 36:115–140.

Schalock, Del
1979   Research on teacher selection. In *Review of Research in Education* 7:364–417.

Schlechty, Phillip C., and Vance, Victor S.
1983   Recruitment, selection, and retention: The shape of the teaching force. *Elementary School Journal* 83(4):469–487.

Schneider, W., and Shiffrin, R. M.
1977   Controlled and automatic human information processing: I. Detection, search, and attention. *Psychological Review* 84:1–66.

Schoenfeld, Alan H.
1982   Measures of problem-solving performance and of problem-solving instruction. *Journal for Research in Mathematics Education* 13:31–49.

Schvaneveldt, R. W., Durso, F. T., Goldsmith, T. E., Breen, T. J., and Cooke, N. M.
1985   Measuring the structure of expertise. *International Journal of Man-Machine Studies* 23:699–728.

Schwartz, Judah L.
1983   *The Semantic Calculator User's Manual.* Newton, Mass.: Education Development Center, Inc.

Schwartz, Judah L., and Yerushalmy, Michal
1985   *The Geometric Supposer.* Apple II computer programs developed by Education Development Center, Inc. Pleasantville, N.Y.: Sunburst Communications, Inc.

Seligman, M. E. P.
1975   *Helplessness.* San Francisco: Freeman.

Shavelson, Richard J.
1972   Some aspects of the correspondence between content structure and cognitive structure in physics instruction. *Journal of Educational Psychology* 63:225–234.
1974   Methods for examining representations of a subject matter structure in a student's memory. *Journal of Research in Science Teaching* 11:231–249.
1985   The Measurement of Cognitive Structure. Paper presented at the annual meeting of the American Educational Research Association, Chicago.

Shiffrin, R. M., and Schneider, W.
1977   Controlled and automatic human information processing: II. Perceptual learning, automatic attending, and a general theory. *Psychological Review* 84:127–190.

Shymanski, James A., Kyle, William C., Jr., and Alport, Jennifer M.
1983   The effects of new science curricula on student performance. *Journal of Research in Science Teaching* 20(5):387–404.

Simon, H. A.
1974   How big is a chunk? *Science* 183:482–488.

Simon, H. A., and Chase W. G.
1973   Skill in chess. *American Scientist* 61:394–403.

Sizer, Theodore
1984   *Horace's Compromise: The Dilemma of the American High School.* Boston: Houghton Mifflin.

Skinner, B. F.
1953   *Science and Human Behavior.* New York: Macmillan.
1968   *The Technology of Teaching.* New York: Appleton-Century-Crofts.

Sleeman, D., and Brown, J. S., eds.
1982   *Intelligent Tutoring Systems.* New York: Academic Press.

South Carolina Department of Education
1986   *State Science Objectives: Grades 1–8.* Columbia: The South Carolina Department of Education.

Stake, Robert E., and Easley, Jack A., Jr.
1978   *Case Studies in Science Education.* NSF SE-78-74. Available from the U.S. Government Printing Office. Washington, D.C.: National Science Foundation.

Stevens, A., Collins, A., and Goldin, S. E.
1979   Misconceptions in students' understanding. *International Journal of Man-Machine Studies* 11:145–156.

Stevenson, Harold W., Lee, Shin-Ying, and Stigler, James W.
1986   Mathematics achievement of Chinese, Japanese, and American children. *Science* 231:693–699.

Stipek, D. J., and Weisz, J. R.
1981   Perceived personal control and academic achievement. *Review of Educational Research* 51(1):101–137.

Summers, Anita A., and Wolfe, Barbara L.
1977   Do schools make a difference? *American Economic Review* 67(4):639–652.

Suydam, Marilyn N.
1984   *Assessing Achievement Across the States: Mathematical Strengths and Weaknesses.* Report No. SE 045 289. Columbus, Ohio: ERIC Clearinghouse for Science, Mathematics and Environmental Education.

Taylor, John R.
1984   A physicist's view of school science. *Outlook* 51(Spring):20–35.

Thorndike, Edward L.
1932   *The Fundamentals of Learning.* New York: Teachers College, Columbia University. Reprinted in 1966 as *Human Learning.* Cambridge, Mass.: MIT Press.

Toffler, Alvin
1980   *The Third Wave.* New York: Morrow.

Traub, R. E., and Fisher, C. W.
1977   On the equivalence of constructed-response and multiple-choice tests. *Applied Psychological Measurement* 3:355–369.

Travers, Kenneth J., McKnight, Curtis C., and Dossey, John A.
1985   Mathematics achievement in U.S. high schools from an international perspective. *Bulletin.* National Association of Secondary School Principals, November 1985:55–63.

U.S. Department of Health, Education, and Welfare
1969   *Toward a Social Report.* U.S. Department of Health, Education, and Welfare. Available from the U.S. Government Printing Office. Washington, D.C.: U.S. Department of Health, Education, and Welfare.

U.S. General Accounting Office
1984   *New Directions for Federal Programs to Aid Mathematics and Science Teaching.* Report No. GAO/PEMD-84-5. Washington, D.C.: U.S. General Accounting Office.

van Dijk, T., and Kintsch, W.
1984   *Strategies of Discourse Comprehension.* New York: Academic Press.

Vernon, P. E.
1962   The determinants of reading comprehension. *Educational and Psychological Measurement* 22:269–286.

Virginia Department of Education
- 1986 *Science Education Program Assessment Model: Resource Guide.* Available from Virginia Department of Education, Richmond, Va.

Walberg, Herbert J.
- 1984 Improving the productivity of America's schools. *Educational Leadership* 41(8):19-30.

Walberg, Herbert J., and Rasher, Sue Pinzar
- 1986 Synthesis of research on teaching. In *Third Handbook of Research on Teaching,* edited by Merlin C. Wittrock. New York: Macmillan.

Walberg, Herbert J., Fraser, Barry S., and Welch, Wayne W.
- 1986 A test of a model of educational productivity among senior high school students. *Journal of Educational Research* 79(3):133-139.

Walker, Decker F., and Schaffarzick, Jon
- 1974 Comparing curricula. *Journal of Educational Research* 44(Winter):83-111.

Ward, W. C.
- 1982 A comparison of free-response and multiple-choice forms of verbal aptitude tests. *Applied Psychological Measurement* 6:1-12.

Ward, W. C., Frederiksen, N., and Carlson, S.
- 1980 Construct validity of free-response and multiple-choice versions of a test. *Journal of Educational Measurement* 17:11-29.

Watts, D. M., and Gilbert, J. K.
- 1983 Enigmas in school science: Students' conceptions for scientifically associated words. *Research in Science and Technology Education* 1(2):161-171.

Webb, N. M., Herman, J. L., and Cabello, B.
- 1986 Diagnosing students' errors from their response selections in language arts. *Journal of Educational Measurement* 23:163-170.

Weber, M. B.
- 1978 The effect of learning environment on learner involvement and achievement. *Journal of Teacher Education* 28(6):81-85.

Weiss, Iris S.
- 1978 *Report of the 1977 National Survey of Science, Mathematics and Social Studies Education.* Prepared for the National Science Foundation. Supt. of Doc. No. 083-000-00364-0. Available from the U.S. Government Printing Office. Washington, D.C.: National Science Foundation.

Weiner, B.
- 1979 A theory of motivation for some classroom experiences. *Journal of Educational Psychology* 71:3-25.

Welch, Wayne W.
- 1983 Research in science education: Review and recommendations. In *Teacher Shortage in Science and Mathematics: Myths, Realities and Research,* edited by John L. Taylor. Washington, D.C.: National Institute of Education.
- 1984 A science-based approach to science learning. In *Research Within Reach: Science Education,* edited by David Holdzkom and Pamela Lutz. Washington, D.C.: National Science Teachers Association.
- 1985 Measuring Student Performance in Science: Trends and Issues. Unpublished paper. Available from the Institute for Student Assessment and Evaluation. University of Florida, Gainesville.

Welch, Wayne W., Anderson, Ronald E., and Harris, Linda J.
1982 The effects of school on mathematics achievement. *American Educational Research Journal* 19(1):145-153.

White, Barbara Y.
1983 Designing computer games to help physics students understand Newton's laws of motion. *Cognition and Instruction* 1:69-108.

Wildavsky, A.
1979 *Speaking Truth to Power: The Art and Craft of Policy Analysis.* Boston: Little, Brown.

Wiley, David E., and Harnischfeger, Annegret
1974 Explosion of a myth: Quantity of schooling and exposure to instruction, major educational vehicles. *Educational Researcher* 3(4):7-12.

Wilks, S. S.
1962 *Mathematical Statistics.* New York: John Wiley & Sons.

Willson, V. L.
1983 A meta-analysis of the relationship between science achievement and science attitudes: Kindergarten through college. *Journal of Research in Science Teaching* 20(9):839-850.

Wise, Arthur E., Darling-Hammond, Linda, and Berry, Barnett
1987 *Effective Teacher Selection: From Recruitment to Retention.* Report No. R-3562-NIE/CSTP. Santa Monica, Calif.: Rand Corporation.

Wolf, Richard M.
1977 *Achievement in America.* New York: Teachers College Press.

Womer, F. B.
1981 State-level testing: Where we have been may not tell us where we are going. In *New Directions for Testing and Measurement: Testing in the States: Beyond Accountability*, edited by D. Carlson, 1-12. San Francisco: Jossey-Bass.

Wooton, William
1965 *SMSG: The Making of a Curriculum.* New Haven: Yale University Press.

Zabalza, Antone, Turnbull, Philip, and Williams, Gareth
1979 *The Economics of Teachers Supply.* Cambridge, England: Cambridge University Press.

Zarinnia, E. Anne, and Romberg, Thomas A.
1986 A new world view and its impact on school mathematics. In *The Monitoring of School Mathematics: Background Papers*, edited by Thomas A. Romberg and Deborah M. Stewart. Vol. 1. Madison: Wisconsin Center for Education Research, University of Wisconsin.

Zarkin, Gary
1985 Occupational choice: An application to the market for public school teachers. *Quarterly Journal of Economics* 100(May):409-416.

Appendix A

Colloquium on Indicators of Precollege Science and Mathematics Education

NOVEMBER 7–9, 1985
Gainesville, Florida

PARTICIPANTS

Lyle V. Jones, *Chair pro tempe*

Testing Group

Norman O. Frederiksen (*Chair*), Director of Psychological Studies, Educational Testing Service*
Charles Anderson, School of Education, Michigan State University
Richard Burton, Cognitive Scientist, Xerox Palo Alto Research Center
Jefferson Davis, Jr., Department of Chemistry, University of South Florida
Pascal D. Forgione, Jr., Chief, Office of Research and Evaluation, Connecticut State Department of Education
Lyle V. Jones, Director, L. L. Thurstone Psychometrics Laboratory, University of North Carolina*
Patrick Kyllonen, Department of Educational Psychology, University of Georgia

---

*Member, Committee on Indicators of Precollege Science and Mathematics Education

Margaret McMeekin, M. K. Rawlings Elementary School, Gainesville, Florida
Ina Mullis, National Assessment of Educational Progress, Educational Testing Service
Jerome Pine, Department of Physics, California Institute of Technology*
Susan Zoltewicz, Chemistry Teacher, Eastside High School, Gainesville, Florida

Curriculum Group

Marshall S. Smith (*Chair*), Dean of Education, Stanford University* (formerly at University of Wisconsin)
Shirley Hill, College of Education (Mathematics), University of Missouri at Kansas City
Edward J. Kormondy, Provost and Vice President for Academic Affairs, California State University, Los Angeles
William T. Lippincott, Director, Institute of Chemical Education, University of Wisconsin*
Wayne Neuburger, Director of Assessment and Evaluation, Oregon Department of Education
Graham Orpwood, Project Officer, Science and Education Study, Science Council of Canada
Andrew Porter, School of Education, Michigan State University
Ramsay Selden, Center for Educational Assessment, Council of Chief State School Officers

Teacher Effectiveness Group

Richard J. Murnane (*Chair*), School of Education (Economics), Harvard University*
J. Myron Atkin, School of Education, Stanford University
Mary Lynn Boscardin, Department of Special Education, University of Massachusetts
Herbert Clemens, Department of Mathematics, University of Utah
Douglas Grouws, Department of Curriculum and Instruction, University of Missouri
Magdalene Lampert, School of Education, Michigan State University

---

*Member, Committee on Indicators of Precollege Science and Mathematics Education

George Miller, Department of Chemistry, University of California, Irvine*
Jeannie Oakes, Education and Human Resources Program, The Rand Corporation
David C. Smith, Dean, College of Education, University of Florida

Science Attitudes/Motivation Group

Mary Budd Rowe (*Chair*), College of Education (Science Education), University of Florida*
Michael Addison, Science Teacher, Buchholz High School, Gainesville, Florida
Robert A. Bernoff, Executive Officer, Pennsylvania State University–Ogontz
James P. Connell, Graduate School of Education and Human Development, University of Rochester
C. Thomas Kerins, Manager of Evaluation and Assessment, Illinois State Board of Education*
Hugh Munby, Faculty of Education, Queen's University, Ontario
Steven Nowicki, Department of Psychology, Emory University
Richard L. Scheaffer, Department of Statistics, University of Florida
Samuel Sebesta, School of Education, University of Washington
Dorothy Turner, Littlewood Elementary School, Gainesville, Florida
Wayne W. Welch, Department of Educational Psychology, University of Minnesota*

Science Literacy Group

F. James Rutherford (*Chair*), Science and Technology Education, American Association for the Advancement of Science
Roger Blumberg, Science Education, Columbia University
James Connor, Department of Science Education, New York University
Alice B. Fulton, Department of Biochemistry, University of Iowa*
David Hawkins, Distinguished Professor, Emeritus, University of Colorado
Phillip Jackson, Department of Education, University of Chicago
Daniel Metlay, Department of Political Science, Massachusetts Institute of Technology

---

*Member, Committee on Indicators of Precollege Science and Mathematics Education

Harold Nisselson, Senior Statistical Advisor, Westat, Inc.*
Parker Small, Professor of Immunology, Medical Microbiology, and Pediatrics, University of Florida

Financial Indicators Group (Special Session)

Kern Alexander (*Chair*), Institute for Educational Finance, University of Florida
1-2 members from each of the other groups

Observers

Richard Berry, Science and Engineering Education, National Science Foundation
Mary Kiely, Program Associate, Carnegie Corporation of New York
Audrey Pendleton, Senior Associate, Office of Educational Research and Improvement

---

*Member, Committee on Indicators of Precollege Science and Mathematics Education

Appendix B

Review of Science Content in Selected Student Achievement Tests

Given the many criticisms of achievement tests, the committee wished to have better information on the quality of the science content of frequently used tests to assess student achievement in science. At a time when achievement test scores have frequently been cited as evidence of declining educational quality in schools, a review of the subject content in science tests appeared to be a potentially useful and important step toward the committee's formulation of recommendations on how to improve indicators of the condition of science and mathematics education.

Two objectives of our review distinguish it from other test reviews. First, we were concerned only with the science content of tests, not the statistical reliability or discriminating power of the items or the test. Second, the review was not designed to produce an evaluation of any particular test or type of test—instead, it was designed to provide information on the quality of the science content found in a variety of achievement tests. Rather than reviewing just one test, reviewers assessed and compared several tests to develop a general picture of the state of science content in achievement tests.

Two primary criteria were used in selecting tests for review: (1) tests of national importance due to the way their results are being used or because they serve as models for other tests and (2) tests that illustrate major variations in purpose and approach so as to provide for a broad assessment of the science content being tested and allow for examination of any difference in content by test purpose. To keep the size of the project within manageable bounds and to provide some test comparability, the age/grade level was limited. Nine tests were selected for review including nationally used norm-referenced tests, state curriculum-based tests, and national and international assessment tests. Table B-1 provides a list of the tests.

A multidisciplinary panel of 12 scientists and science teachers was selected to conduct the test review (the panel list appears at the end of Appendix B; two of the individuals listed did not review tests individually but wrote general comments). The panel was constituted so as to combine the perspectives of people from different science fields and different professional positions—college professors, research scientists, and secondary school science teachers. Two specialists in cognitive learning processes who have studied science education and testing were also included in the group.

The test review process was planned to have two stages, individual ratings of tests and a subsequent meeting for group discussion. In the first stage, each test was reviewed by one physical scientist, one life scientist, one cognitive scientist, two teachers, and a sixth reviewer from one of these categories to allow for comparisons of test reviews by type of reviewer. The reviewers rated each individual test item and then analyzed each test as a whole. The test items were rated according to two criteria: (1) importance—the reviewer's assessment of how important the knowledge being tapped is for a student and (2) adequacy—how adequately the item tests that knowledge, given the purpose of the test. Several patterns emerged from the ratings:

- The scientists in the group were more critical of the science content of the tests than were most of the teachers. One explanation for this difference might be that scientists expect greater quality of science content in the tests than do teachers. Another possible explanation is that the teachers are more familiar with these tests, as well as with other achievement tests, and do not see as many problems in the actual use of the items.
- The science teachers were more critical of the norm-referenced tests than of the other types of tests. The teacher reviewers seemed to find more problems with this type of science test than with the tests used for national assessments or curriculum-based tests.
- There appeared to be a relationship between the science field and the item ratings of a reviewer. The two biologists rated the New York State Regents biology test lower than any other reviewers.

APPENDIX B

TABLE B-1 Science Tests Selected for Review and Average Student Scores on Each Test

Test	Average Percentage Correct	Number of Items	Comments
High School and Beyond (HSB)	46.5	20	Science portion of test; score is for national sample of 1980 10th-grade students
National Assessment of Educational Progress: 13-year-olds (NAEP-13)	52.4	77	Scores are for 1981 test given to a national sample; no scores were available for 1985-1986 test that was reviewed
National Assessment of Educational Progress: 17-year-olds (NAEP-17)	60.0	56	
California Assessment Program (CAP)	53.8	—	1984-1985 field test of 1,650 items given to over 10,000 California 8th-grade students; average score over six different categories of questions
Comprehensive Tests of Skills (CTBS)	52.5	40	1982 norm for end of 9th-grade score at 50th percentile of all students taking the test
Tests of Achievement and Proficiency (TAP)	53.3	60	1982 norm for spring 9th-grade score at 50th percentile of all students taking the test; no norm was available for 1985-1986 test that was reviewed
International Association for the Evaluation of Education Achievement (IEA)	(M) 64.7 (F) 58.3	90	1983 test; score is for U.S. sample of 9th-grade students
New York State Regents: Science (NYSR-ES)	77.1	105	65 percent correct is Earth minimum passing score; 79.8 percent of 37,175 students passed in June 1984
New York State Regents: Biology (NYSR-BIO)	74.7	103	65 percent correct is minimum passing score; 72.8 percent of 114,068 students passed in June 1984

Possibly the biologists could find more problems with the items due to greater knowledge and familiarity with the current state of the field.

The second stage of the review process consisted of a group discussion of the nine tests among the group of reviewers. For this purpose, a two-day meeting was held at the National Academy of Sciences. The meeting had three components: discussion of the item ratings and qualitative test analyses by the reviewers, identification of common findings concerning the science content in the nine tests, and outlining of the characteristics of good science tests. The major outcome of the meeting was the development of some qualitative conclusions on the current state of science testing and suggested improvements that should be pursued.

Differences in average ratings between the tests were relatively small compared with the variability between the reviewers. However, the science test reviewers reached four general conclusions:
- The nine science achievement tests typically cover broad content areas, and the content is generally appropriate for the intended grade level; however, a majority of the tests are weak in testing core science concepts and depth of student knowledge.
- Five to ten percent of the items on each test include inaccurate or misleading science statements that decrease the usefulness of the test results.
- The tests vary widely in the quality and balance of items intended to test different types of skills, that is, factual knowledge, concepts, science processes, reasoning, and problem solving.
- The format, language, and structure of science tests strongly affect the usefulness of test results for educational and assessment purposes.

Based on its discussions, the group identified characteristics of high-quality science tests according to testing purpose.

For national, state, or local assessment:
- Assessment items should be based on a sampling of the ideal or desired curriculum in the subject area.
- Items should focus on central concepts for the course or grade level.
- Given the identification of the core subject matter to be covered, the test should be designed from a matrix of desired learning

APPENDIX B   179

objectives, consisting of elements of the subject knowledge base classified by the types of desired skills.
• A few items should offer new ways of thinking about a concept or solving a problem and provide topics for teachers to use in subsequent instruction.
• Test results should be reported to local test users, for example, administrators, teachers, parents, and students, in relation to the matrix of objectives so as to increase the educational use of assessment results.

For rank-ordering of student performance:
• The test should be designed to assess knowledge that is closely related to the reason for the ranking.
• There should be less stratification of students by test performance, because often it is based on misuses of small differences in test results.

For diagnosis and guiding instruction:
• Diagnostic tests should be written with a real-world orientation, that is, without subject-specific jargon and terminology, and they should include samples of different kinds of science experience the student may have had and science ideas the student may understand.
• Time allowed for conducting a diagnostic test is an important design variable, because some students do not perform well under time constraints.
• Test results should be reviewed item by item rather than as an overall test score. Since a test can sample only a limited portion of the total knowledge of a student, performance on individual items rather than on the test as a whole should be used to assess student knowledge for purposes of diagnosis.
• As they employ diagnostic tests, teachers should prepare administrators, parents, and students to understand the meaning of test results and carefully explain how they will be used.
• The use of achievement tests for diagnosis and improving instruction could be advanced if testing were less dependent on methods involving only paper and pencil. Alternative technologies for diagnostic testing in science need to be further developed.
• The results of research in cognitive science and other educational research should be used in test development.

The group also made the following suggestions to avoid the misuse of tests:

- Results from tests constructed for one purpose, for example, rank-ordering of student performance, should not be used for a quite different purpose, for example, assessing instructional quality.
- School or classroom average test scores should not be applied to individuals, and individual test scores should not be interpreted as a rating or ranking of the persons, but only of performance on a test that assesses specific skills.
- Test results or tests of the kind reviewed should not be used as the major force driving curriculum and instruction.

SCIENCE TEST REVIEW PANEL

Marshall S. Smith (*Chair*), Stanford University (education, measurement, and evaluation)

Andrea diSessa, University of California, Berkeley (cognitive science)

Rachel Egan, Orchard Ridge Middle School, Madison, Wisconsin (science teacher: eighth grade)

Joyce Gellhorn Greene, Boulder High School, Boulder, Colorado (science teacher: biology)

Henry Heikkinen, University of Maryland (chemistry)

Jack Lochhead, University of Massachusetts (cognitive science)

Lucy McCorkle, Cardozo High School, Washington, D.C. (science teacher: chemistry)

José Mestre, University of Massachusetts (physics/cognitive science)

James Minstrell, Mercer Island High School, Mercer Island, Washington (science teacher: physics)

Philip Morrison, Massachusetts Institute of Technology (physics)

Phylis Morrison, Cambridge, Massachusetts (elementary science teacher)

Wayne Moyer, Franklin Institute Science Museum and Planetarium, Philadelphia, Pennsylvania (biology)

David Policansky, Commission on Life Sciences, National Research Council (biology)

Appendix C

Summaries of Meetings with Representatives of State and Local Education Agencies

SUMMARY OF MEETING WITH REPRESENTATIVES
OF STATE EDUCATION AGENCIES

APRIL 16, 1986
San Francisco

The purpose of the meeting between representatives of state education agencies (see the list of participants below) and members of the committee was to provide an opportunity to discuss mutual interests concerning the assessment of the quality of science and mathematics education. The committee presented some preliminary ideas on six indicator areas and asked for reactions from the state representatives as well as discussion of additional concerns they wished to raise.

Committee members summarized draft statements that had been circulated before the meeting on assessment of the quality of the curriculum, teacher effectiveness, student learning, investment of resources, student attitudes and motivation, and scientific literacy. Following each presentation, the state representatives commented on the feasibility and desirability of the suggested indicators and proposed other indicators that might be considered. The comments and discussion are summarized below under each indicator area.

Quality of the Curriculum

- A framework for assessing the quantity and quality of curriculum content in each subject area would be very useful and desirable at this time. The response to the construction of such frameworks would be positive on the part of those concerned with educational improvement because more direction is needed on priorities in curriculum content. In that connection, it might be worthwhile to review curricular frameworks used in other nations, for example, West Germany, France, Japan, and Great Britain.
- The coherence of the curriculum across grade levels is important. The quality and quantity of subject matter to which a student is exposed should not be assessed within a grade level or course only, but over a reasonable period of schooling, e.g., primary grades. In that way there could be some latitude regarding the sequencing of units, for example, a core topic might be taught in either third or fourth grade. The framework idea might lead to a useful "national grid" of science and mathematics subject matter that identifies key concepts and processes to be included in the curriculum but without specifying the exact placement.
- It may be difficult to capture quality in science curricula through the framework concept as outlined by the committee, because there are different approaches and philosophies that prevail in the teaching of science, often having to do with the sequencing of topics. But if the sequence or grade level for introducing a particular topic is not highly specified in the framework, teaching approach may not be an issue.
- It is critical to maintain the distinction between a "national" curriculum framework and a "federal" curriulum framework—that is, between a set of guidelines developed by one or more nationally recognized groups and a prescribed course of studies mandated by a central authority. A national framework could have an important function in making possible comparisons and evaluations of the content of various state assessment tests and commercial achievement tests in specific subjects.
- South Carolina has developed a science curriculum framework for grades 1 through 8 that may be of use to the committee as an example and for comparison with other frameworks. South Carolina would have found the product of a national effort, such as the current one by the committee, valuable when they were working on a state framework. California also has developed a science curriculum framework in conjunction with the new state science assessment test

for the eighth grade. New York is an example of a state with a science curriculum for grades K–12.
- An additional perspective on curriculum assessment could be offered by people who are external to the education system but who have certain expectations of students with respect to their science and mathematics education. Groups to be consulted might include employers, college-level scientists, and scientists in industry, all of whom are influential in the determination of the intended curriculum, i.e., what the schools should be held responsible for in science and mathematics.

Teacher Effectiveness

- Some measure of subject matter preparation should continue to be considered an indicator of teacher effectiveness. Agreement on specifics may be difficult, however, since no satisfactory determination may be possible at this time of optimal preparation for teaching a subject at a given grade level or teaching a particular course.
- Even if the relationship between subject matter preparation and effective teaching of a subject were better understood, there would still be problems with current teacher tests.
— Tests for elementary teachers lack science content altogether; typically, they are dominated by questions on general pedagogy. The low expectation for instruction in science at the elementary level may be a contributory factor, as may be the absence of any agreement as to what the science content of the elementary school curriculum should be even when science is being taught.
— A more general criticism of teacher testing is the extent to which coaching can and has been used to improve test scores, thus decreasing a test's validity. One approach, used in South Carolina, is to disseminate test specifications that indicate the areas to be tested, but not to distribute or coach on sample test items.
— The impact of teacher tests on preservice and in-service education must be considered, analogous to the impact of student achievement tests on the school curriculum. The implication is not to do away with teacher testing, but to improve the tests so that they assess important rather than trivial knowledge and process skills, again analogous to the improvement needed in student tests. If there were national curriculum frameworks for science and mathematics, they could guide the content of the teacher tests as well as of student tests.

- How a teacher actually delivers the curriculum to the students importantly affects what subject-matter content they are likely to learn. Therefore, the quality of curriculum delivery needs to be assessed, and appropriate indicators need to be developed. Given the disparity in science and mathematics learning among different student groups, the indicators must be sensitive to variations in delivery according to the range of students in a classroom or a school.
- At present, the two methods used to assess curriculum delivery are classroom observation and "opportunity-to-learn" questionnaires administered to teachers and older students, as in the IEA and NAEP assessments. Although costly, observation should be included as a recommended method. The Tennessee assessment of teacher effectiveness for the state's career ladder program included three outside observers. An important benefit of the observations was that teachers were able to reflect on their behavior and techniques in the classroom. Items that differentiated outstanding teachers in Tennessee were the extent of planning, use of a variety rather than just one or two teaching strategies, and instruction in higher-order thinking skills. It is also important to observe the teaching of a range of students, not just the better students in science and mathematics.
- The notion of adding intellectual curiosity to the other two factors that make for teacher effectiveness (subject-matter knowledge and ability to get knowledge and intellectual curiosity across to students) is important. The difficulty of assessing this factor should not deter the committee from including it; rather, work on developing useful indicators of intellectual curiosity needs to be encouraged.
- Observation should include some higher-inference items, particularly to assess adequately the teaching of higher-order skills. Many observation instruments concentrate on lower-inference items because observers can be trained more easily, and they yield higher reliability.

Assessment of Learning

- The provisional draft [an early version of Chapter 4] developed during the committee's workshop on learning assessment provides an exciting, forward-looking statement on cognitive processes and testing. It is useful at this time when states are considering possibilities for computerized testing.

- Statewide tests can and do have a great impact on curriculum and teaching. The committee could provide very useful advice and models on how to measure higher-order skills through statewide tests.
- Matrix sampling is a possible approach to testing higher-order skills through new testing methods. However, some states have mandated individual testing of all students. There are two primary reasons for individual rather matrix-sample testing: (1) students and parents want individual test scores for external uses and (2) comparisons between schools are more difficult with matrix sampling unless there is a sufficient number of students tested in each school. California uses matrix-sample testing and has obtained reliable school comparisons with testing 30 students per school. Florida, Tennessee, and Virginia have used matrix sampling of students with a regional study of eleventh grade reading that will produce state comparisons.
- Matrix sampling can involve selection of different combinations of items for each student, and, if desired, all students can be tested. This approach increases the content covered and tested, a considerable advantage for assessing the quality of programs in a school. Matrix sampling places more pressure on the school staff and decreases reliance on student variables rather than school variables to explain success or failure on the tests.
- Most current achievement tests do not test what an individual student knows, since they sample only a small portion of the curriculum. Computerized methods of testing would allow much greater coverage of what a student knows and does not know and thus permit teaching to student deficiencies.
- Nevertheless, the highest priority for developing and using test information should be to assess the effectiveness of a school program or curriculum. Although individualized programs for students are often discussed, it is unrealistic to place priority for use of testing on individual student diagnosis and design of individualized instruction.
- The item bank concept is difficult to put into effective practice. Access to the items is crucial, and that will entail a good deal of careful planning. A number of states have item banks for science, such as Oregon, Minnesota, and North Carolina, and other states, including Florida, are considering item banks. Some of the current item banks are not well utilized. A national library of items, like the one the committee outlined, would provide a framework for classifying items that are compiled by states. Another item bank may not be needed, but a conceptual model for use of items is needed.

- Models of good items that assess process and higher-order skills are also urgently needed. An enormous amount of scientific expertise is necessary in test development and validation. There is large potential for misinformation in poor item stems and distractors, and too many item reviewers are not expert in the areas of science for the items they are reviewing. The committee could serve an important role in developing an item library that concentrated on creating high-quality items and on models for use of the items.
- It is difficult to move away from such simple quantitative indicators as test scores or the "science dropout rate" toward qualitative indicators that would report more information. One view expressed was that multiple test scores would be better than one score. Another view was that it might be possible to construct a scale of science learning that could be compared with desired curriculum outcomes. Such scales do exist for reading and mathematics. If qualitative indicators are to be reported to state and local policy makers to give greater depth of information, a common "language" for qualitative indicators would need to be specified, i.e., consensus would need to be established on the meaning and interpretation of words used to express the indicators.
- Experts that develop and recommend indicators to policy makers should have a clear idea of what is important to know and the purpose of the information. Parsimony with indicators is crucial. Much of the data currently collected by state education agencies is not used.

Use of Resources

- Indicators for resource use at the local level should focus on availability of resources in the classroom and resource use from the teacher's perspective. It is too difficult to interpret such centrally collected measures as full-time-equivalent staff with respect to programmatic significance, i.e., resource investment in, say, physics or mathematics.
- States are quite aware of the decline in federal resources for science. For example, the NDEA grants (in the 1960s) were the last major federal funds for equipment and supplies in science. The waxing and waning of federal resources for science (and other programmatic areas) should be tracked.
- State agencies generally have not committed funds for resupplying equipment and materials for science, even though these are urgently needed in districts. Often, other school funding priorities

take precedence, such as raising teacher salaries. Since 85 percent of the typical school budget is allocated to staff salaries and benefits, there is little wiggle room in the budget. In any case, states prefer to let local school districts allocate funds by program, thus moving competition for funding to the local level.
• Recent changes in state graduation requirements in science and mathematics are having important impacts on local resources. Requirements that each school offer advanced science courses are being instituted in a number of districts and states; such courses are especially costly to teach and may draw resources (e.g., the best teachers) from other science instruction.

Student Attitudes and Motivation

• The committee's statement focused mainly on indicators of scientific attitudes possessed by students. Another approach is to assess student attitudes toward science classes, science teachers, or the scientific disciplines themselves. For example, the NAEP 1982 survey revealed that only 35 percent of students think their teachers like science. Student images of science and scientists may be important factors in motivating students to learn science and in career decisions.
• Some states include items on student attitudes in their assessments. For example, the California eighth-grade science test includes 30 such items; initial results were made available in August 1986. Further information on what states are doing in this area should be available from the UCLA Center on Evaluation, which has reviewed state assessment instruments, including attitude items.

General Science Literacy

• The committee's statement on scientific habits of mind bears some similarity to its statement on student attitudes and motivation, particularly with respect to learning to think about natural phenomena as do scientists.
• The committee's perspective on science literacy is an excellent general statement of the role and importance of science education; it provides a good rationale for science preparation for all citizens, not just preparation of scientists. The committee should consider introducing its report with this statement.

Participants: State Education Agencies

Dale Carlson, Director, California Assessment Program, California Department of Education

David Donovan, Assistant Superintendent for Technical Assistance, Michigan Department of Education

Janice Earle, Maryland State Department of Education

Gordon Ensign, Supervisor of Testing and Evaluation, Washington Superintendent of Public Instruction

Pascal D. Forgione, Jr., Office of Research and Evaluation, Connecticut Department of Education

Steven Koffler, Bureau of Cognitive Skills, New Jersey Department of Education

Windsor Lott, Director, Division of Education Testing, New York State Department of Education

George Malo, Tennessee Department of Education

Wayne Neuburger, Director, Assessment and Evaluation, Oregon State Department of Education

Paul Prowda, Office of Research and Evaluation, Connecticut Department of Education

Edward Roeber, Michigan Department of Education

Paul Sandifer, South Carolina State Department of Education

Ramsay Selden, Director, State Education Assessment Center, Council of Chief State School Officers

Janice Smith, Assessment, Evaluation, and Testing, Florida Department of Education

Zack Taylor, Science Unit, California Department of Education

Suzanne Triplett, State Education Assessment Center, Council of Chief State School Officers

Marvin Veselka, Assistant Commissioner of Assessment, Texas Education Agency

SUMMARY OF MEETING WITH REPRESENTATIVES OF LOCAL SCHOOL DISTRICTS

JUNE 6, 1986
Washington, D.C.

The purpose of the meeting between representatives of local school districts (see the list of participants below) and members of the committee was to provide an opportunity to discuss mutual interests concerning the assessment of the quality of science and mathematics education. The committee presented some preliminary ideas on six indicator areas and asked for reactions from the local representatives as well as discussion of additional concerns they wished to raise.

Committee members summarized draft statements that had been circulated before the meeting on assessment of teacher effectiveness, the quality of the curriculum, student learning, investment of resources, scientific literacy, and student attitudes and motivation. Following each presentation, the local representatives commented on the feasibility and desirability of the suggested indicators and proposed other indicators that might be considered. The comments and discussion are summarized below under each indicator area.

Teacher Effectiveness

Indicators of teacher effectiveness need to be tied to clearly stated assumptions about the goals of science and mathematics education; e.g.: student achievement test scores need to be raised; the number of college students majoring in scientific fields needs to be increased; or the overall science literacy of all 18-year-olds needs to be raised. These goals are not necessarily mutually exclusive, but they may require different teacher competencies.

Possible Indicators

- Teacher effectiveness is not a unitary variable that can be measured along a single dimension. It needs to be assessed in the context of specific subject matter, at particular grade levels, and with respect to groups of students with different levels of ability and coming from different socioeconomic backgrounds.
- Related to the first comment is the need to appraise the effectiveness of a teacher in organizing and presenting instruction to meet the needs of students. Twenty years ago, when students were

differently motivated, it may have been appropriate to emphasize subject-matter knowledge of teachers as a prime requisite for teaching. The needs and backgrounds of many students are more varied today; teachers must have empathy and understanding as well as subject-matter knowledge in order to teach most students. Variations in teacher effectiveness, however, should not be explained away by the characteristics of students—i.e., the background and ability level of students should not be used as an excuse for ineffectiveness of the teacher or the school.

- Also related to the first comment, any indicator of intellectual curiosity should use different measures for elementary and secondary teachers, given different responsibilities and expectations for teachers at each of these levels. Measures might also differ for teachers of advanced placement versus basic skills classes, although having different standards for teachers may be a subtle form of failing to hold teachers responsible for low student performance.
- The attitudinal or motivational aspect of teacher effectiveness should be discussed by the committee as a potential indicator, analogous to consideration of indicators of student attitude and motivation.
- Some local school districts prefer an outcomes-based model for measuring teacher effectiveness, as opposed to assessing teacher characteristics (e.g., intellectual curiosity) or using process measures. An outcomes-based model provides for assessment of the contribution of a teacher to student learning and educational attainment over time, while taking into account the effects of student background and school and teacher characteristics. Several kinds of outcomes measures, in addition to test scores, can be included, for example, graduation and dropout rates, proportion of students going to college, and various honors and awards earned by students. A potential difficulty in implementing this kind of model is the high degree of student mobility between schools, districts, and states.

Use of Indicators

- Any recommendation to test teachers for subject-matter knowledge should specify that test results not be used for evaluating individual teachers, either for entering or advancing in a teaching job. Items asking for demographic information on teachers should be excluded from subject knowledge tests to ensure that the results are used only to assess the overall quality of the teaching staff of a district. If demographic information is collected, some means of ensuring anonymity of responses should be provided. Within these considerations, a test of teachers' minimum level of competence in their subject would be a useful indicator for local school districts.

- A standard for minimum competence in a teacher's subject should be considered a threshold level of competence. Testing of teachers' subject knowledge probably needs to extend slightly beyond the level at which they teach. That is, teachers need to know what a student will be learning at the next level and how instruction at the two levels is related.
- Recommendations for indicators of teacher effectiveness should be accompanied by recommendations on the appropriate level of analysis of the indicators, i.e., individual teacher, school, district, state, nation. This is important for the design of specific measures and the use of indicators.
- The interest of teachers in indicators, as reported by one of the LEA representatives, relates mostly to aspects of their job that they perceive need improvement, e.g., time available for professional development and planning of instruction.

Regarding the committee's work, conflicting views were expressed on the usefulness for constructing indicators of the existing research on teacher effectiveness and school effectiveness. One view was that the committee's report should take account of the main findings coming from this research, even if the indicators recommended by the committee are not necessarily based on the findings. Many districts have designed programs to improve instruction based on school effectiveness research. A second view was that much of the research on school effectiveness and teacher effectiveness is flawed methodologically, and thus the committee need not worry about citing the findings.

Quality of Curriculum

Analogous to teacher effectiveness indicators, recommendations on assessing curriculum quality also need to be tied to assumptions about educational goals, i.e., the expected performance level of students in science and mathematics. Curricular frameworks cannot be constructed nor core concepts specified without knowing what level of knowledge is expected of students—minimum competency, science literacy, or college preparation. If that is its intent, the report should state clearly that the committee's goal is to assess science and mathematics curricula, and learning, for all levels of students.

Possible Indicators

- Frameworks for assessing the quality of curriculum are very important and urgently needed; they would be especially useful if they connect "strands" of curriculum objectives between the grades. A framework or set of core concepts needs to be fairly specific to provide a means of assessing differences between programs and schools. Local districts would like to be able to provide evaluative information of this kind for their curriculum specialists. Given the current state of curriculum development, frameworks are more applicable to the mathematics curriculum than the science curriculum.
- In assessing the quality of the curriculum, factors in addition to the framework or set of core concepts should be considered, including community needs and interests. Frameworks must allow for local variations in the curriculum.
- A potential indicator of the quality of the curriculum in high schools is "holding power"—the extent to which students continue to enroll in courses within a subject area.
- According to some LEA representatives, the proposed method of measuring the "taught curriculum"—through self-reports by teachers—will not produce a valid indicator of the curriculum that is actually taught. Teachers will tend to overreport what they cover, especially if they think their response will be used to evaluate their performance. It was pointed out that self-report measures have been used on previous studies, e.g., the IEA Mathematics Assessment. In that case, when coverage was being tied to student performance, it might have been in the teachers' interest to underreport what topics have been covered. In either case, teacher self-reports may not yield accurate estimates of what is taught in classrooms. Self-reports could be corroborated by random auditing procedures.

Assessment of Learning

- Several points were raised concerning the feasibility of the recommended national library of test items and how it might be implemented. Quality control of the items is a major issue that will need to be resolved. Also, the library should have a method of tracking the use and effectiveness of items, possibly by monitoring which

APPENDIX C 193

items are requested and asking LEAs to return information on their experience with items, including statistical data on scores.
• For science, hands-on assessment items should be included in the materials in the library. NAEP is currently testing out some hands-on items.
• Many larger districts have developed their own criterion-referenced tests because sufficiently comprehensive item banks to allow choices to match curricula were not available. Some districts are using items that were developed for the high schools in Dade County, Florida. Locally developed tests have the advantage of giving teachers a feeling of ownership and involvement in the curriculum and testing process.
• Data obtained from locally developed criterion-referenced tests could be used more extensively for diagnostic purposes with students, comparisons of schools and classes, and analyses of grades that are assigned to students. Local districts and schools need to make better use of existing tests and data for assessment of learning, while development of improved tests and assessment methods continues.

Resources

• Indicators of resources for science and mathematics should be based on actual use, for example, the number of students in a school using the science laboratory and how it is being used. The mere presence of laboratories, or even their availability to the teacher, is not really important. Their value is in the extent and quality of use with students.
• A much more important resource issue than laboratories, facilities, or supplies for science and mathematics is the use of resources for teacher training and teacher development, i.e., preparing teachers to improve their teaching by more effective use of such resources as laboratories.
• Information on resources for science and mathematics could be very valuable, but a major question is how the data should be collected. One option suggested was to use the accreditation process to identify availability and use of resources. However, accreditation is already burdensome for schools and accrediting committees. Moreover, accreditation tends to be based on subjective reviews and assessment rather than collection of quantitative data. The method and organization selected for collecting information on resources is likely to have considerable effects on how the information is used.

- The committee should not ignore the level of federal investments in recommendations for indicators of resources. For example, current initiatives to encourage retraining of teachers for shortage areas have implications for federal policies, and additional funds will be needed.

Scientific Literacy

- A question was raised concerning the possibility of using a composite measure of scientific literacy rather than several different measures, as suggested in the committee's statement. However, a composite measure is likely to mask differences on the several dimensions of scientific literacy discussed in the statement, and the interpretation of separate measures matching these dimensions would be more straightforward and valid.
- The committee's draft statement calls for "flexible" indicators. A better description of the desired attribute might be to call for indicators that are "sensitive to change."
- The committee should consider defining scientific literacy, including aspects of technological literacy, from the perspective of employers. Opinions differ on what constitutes effective education for current and prospective job markets: one view emphasizes knowledge and understanding of technology; another view holds that the basics of science and mathematics are more important, given the rapid changes in technology (e.g., the shift from transistors to microprocessors).
- No matter how technological literacy is defined, it is hardly taught at all at present. Hence, increasing the technological literacy of students would involve high costs for developing appropriate curricula and—even more so—the needed skills and knowledge of teachers.
- Assessment of scientific literacy should include students still in school as well as adults in order to measure change over time, i.e., what people retain of what they have learned during their school years and what new concepts, information, and skills they have acquired.

Student Attitudes and Motivation

- The committee considers student attitudes to be an outcome of instruction in science. However, student attitudes toward science can be strongly affected by the attitudes of peers and adults. In particular, attitudes can be shaped by teachers at a very early point in education. This reinforces the suggestion made above on assessing teacher attitudes and motivation as well as those of students. Intrinsic interest and motivation toward science is needed by teachers for good science teaching, just as it is needed by students for science learning. Student attitudes and motivation might be analyzed in relation to teacher attitudes and motivation.

- It is very important to learn more about the affective component of science and mathematics education. This is particularly important for local school districts at the present time as requirements for the number of science and mathematics courses are being raised in the face of demonstrated low student interest in these subjects. Better information on attitudes and motivation may yield clues as to the reasons why most high school students avoid science and mathematics courses if not forced to take them.

General Suggestions

- The term *precollege* is too narrow, given the goals of science education assumed in the report, i.e., improving science and mathematics education for all students. *Precollege* implies interest only in college-bound students.
- Indicators should provide the capacity for assessing the long-term impact of education on such goals as increasing scientific literacy or increasing interest in science and mathematics, not just immediate results, for example, outcome measures that reflect the goal of raising test scores.
- The scaling of indicators is important. Measures need to be expressed in terms of distribution or range, not simply averages or means.
- The committee should consider recommending that more research on indicators be conducted involving large school districts because many have large, accessible data bases for carrying out research.
- Recommendations for new indicators are likely to require different kinds of evaluation and research on elementary and secondary education than in the past. The development of indicators useful at state and local levels may well affect the current roles and practices of local and state agencies in collecting, analyzing, and using data.

Participants: Local School Districts

Alan Barson, Curriculum Division, Mathematics and Science, School District of Philadelphia
Milton Binns, The Council of Great City Schools
Frances M. Culpepper, Science Coordinator, Atlanta Public Schools
Stephen H. Davidoff, Research and Evaluation, School District of Philadelphia
Steven Frankel, Department of Educational Accountability, Montgomery County Public Schools, Maryland
Joy Frechtling, Department of Educational Accountability, Montgomery County Public Schools, Maryland
LaMarian Hayes-Wallace, Office of Research and Evaluation, Atlanta Public Schools
Paul Hovsepian, Divisional Director, Mathematics and Science, Detroit Public Schools
Sam Husk, Executive Director, The Council of Great City Schools
Joseph P. Linscomb, Office of Associate Superintendent of Instruction, Los Angeles Unified School District
Joy Odom, Coordinator, Secondary Mathematics, Montgomery County Public Schools, Maryland
Joyce Pinkston, Coordinator of Curriculum Development, Memphis City Schools
Harold Pratt, Science Coordinator, Jefferson County Public Schools, Colorado
Kathy Pruett, Director, Research Services, Memphis City Schools
Stuart C. Rankin, Deputy Superintendent, Educational Services, Detroit Public Schools
Thomas Rowan, Coordinator, Elementary Mathematics, Montgomery County Public Schools, Maryland
Nicholas Stayrook, Director, Evaluation Services Department, Seattle Public Schools
Floraline I. Stevens, Director, Research and Evaluation, Los Angeles Unified School District
Gary Thompson, Department of Evaluation Services, Columbus Public Schools
Ray Turner, Assistant Superintendent for Educational Accountability, Dade County Public Schools, Florida
Robert Wright, Secondary Science, Curriculum Specialist, Seattle Public Schools

Appendix D

Current Projects on Indicators

The report of the National Commission on Excellence in Education (1983), the report of the National Science Board Commission on Precollege Science, Mathematics, and Technology Education (1983), and other recent national reports found that better information is needed on the status of education in American schools and recommended that better indicators be developed for measuring change in the quality of education. The committee's first report (Raizen and Jones, 1985) provided an analysis of the current state of indicators for assessing precollege science and mathematics education, and that report reviewed existing data-collection efforts that may contribute to indicators. The report also highlighted the kinds of data and information that should be available to policy makers, but often are not.

Since that report was completed, a number of studies and activities have begun at the national level to fill the gaps and inadequacies in indicators of science and mathematics education. Many of these studies and activities are proceeding or being completed as the committee is writing this report, and since many of the findings and recommendations are likely to complement this report, the objectives, designs, and potential findings of these studies and activities are outlined below.

Development of Indicators from Existing Research and Data Bases

A new synthesis of information was created in direct response to the report of the National Science Board Commission. The board decided to include a chapter on science education in its biennial *Science Indicators* report to Congress. The chapter included in the 1985 *Science Indicators* report (National Science Board, 1985) provides a review of available data that can be used to monitor and assess the quality of science education, such as academic performance of students, academic standards for science and mathematics, adequacy of the curriculum, and qualifications and supply of teachers. The chapter also identifies some problems in answering questions about the quality of science and mathematics education that derive from the failings of existing indicators.

The U.S. Department of Education (1985) also responded to the expressed need for indicators by developing a new report, "Indicators of Education Status and Trends," which was released in 1985. This report identifies 20 indicators for measuring the quality of schooling, using existing national data. Department of Education staff worked with a consortium of 20 national education organizations to select indicators in three categories—outcomes, resources, and context—that would provide the most meaningful and useful measures of educational quality in elementary and secondary schools. More recently, the department released a briefer version (U.S. Department of Education, 1987) containing 17 indicators derived from existing data bases and research studies. The definition of an education indicator, as given in the department's first report, has been frequently cited in other studies and activities on indicators.

A study report by the Rand Corporation (Shavelson et al., 1987) for the National Science Foundation is concerned with the design of a national indicator system for monitoring science and mathematics education. The two major objectives of the study were, first, to consider the benefits of alternative monitoring systems and, second, to determine the feasibility and cost of each alternative system. The study report incorporates findings and recommendations on indicators from the committee's first report as well as recent research and data collection that would contribute to a monitoring system.

The Educational Testing Service is conducting a study for the National Science Foundation to determine the possibility of developing a comprehensive unified data base for science education indicators. The goal of the project is a single data base that will provide current information on the number and quality of students in science and engineering at several stages in their development. The plan is to integrate data from about 25 existing large data bases that have information on demographic characteristics and educational achievement of precollege and college students, such as the National Assessment of Educational Progress, the Scholastic Aptitude Tests (SATs), the High School-and-Beyond Longitudinal Survey, and the Graduate Record Examinations. The integrated data base would be updated annually to provide the capacity for analyses of the status of students in science and engineering from one year to the next. The comprehensive data base would allow the National Science Foundation to answer regularly such questions as the change in quantitative ability of high school seniors from year to year, the expected number of graduates planning to enroll in college science majors, or the science achievement of specific groups of minority students planning to enter teaching.

A study recently completed by the Center for the Study of Evaluation at the University of California, Los Angeles (Burstein et al., no date) explored the feasibility of using existing data collected by the states to construct education indicators. The study was conducted for the U.S. Department of Education in response to questions about using state testing data for state-by-state comparisons of student performance at the national level. The goal of the study was to examine methodological and implementation issues in aggregating data from state testing programs and then to recommend ways of facilitating their use on a national basis. The study included analyses of the current state testing programs, discussion of alternative approaches to linking test results across states to create a common scale, and assessment of the availability of information about schools and students that could be used to construct more valid indicators of achievement.

The Center for Policy Research in Education, funded by the U.S. Department of Education, conducts research on state and local education policy in order to foster educational improvement. The center recently published a guide to indicators (Oakes, 1986) designed to acquaint policy makers with the development, interpretation, and use of education statistics. Topics covered include definition and types of indicators, the use of indicators, indicators in a policy context, and state of the indicator art.

In response to states' interests in better indicators of education, the Council of Chief State School Officers has created a state education assessment center to develop and coordinate an educational indicators system for use by all the states. The center will work with states to develop a common set of indicators, including selection of indicators, identification and improvement of existing data gathered by states, and design of new data bases when necessary. Indicators will be developed in three areas: (1) the context in which education takes place, including the demographics of the population, the resources available, and student descriptors; (2) educational policies and practices, including amount and use of instructional time, the instructional program, preparation and characteristics of teachers, the allocation of resources, and policies on school participation; and (3) educational outcomes, including student achievement, attendance, school completion, and post-school outcomes and attitudes. A special task force, supported by the National Science Foundation, is working on state science and mathematics indicators. The committee has worked closely with the assessment center staff in ensuring that the interests and needs of states for improved indicators are reflected in this report.

Studies to Improve Basic Data Collection

Another study of the National Research Council concerns statistics on supply and demand for precollege science and mathematics teachers. The panel of experts conducting the study was selected in consultation with the Committee on National Statistics and this committee. The panel is currently evaluating models used at national and state levels for estimating and projecting teacher supply and demand; it is also assessing the measures of teacher qualifications used in these models. The goals of the study are to develop a method for constructing a national profile of teachers, to recommend further data collection necessary to provide more adequate information on teacher supply and demand at national and state levels, and to outline improved models for projecting teacher supply and demand and estimating effects of alternative policies. An interim report is available (Panel on Statistics on Supply and Demand for Precollege Science and Mathematics Teachers, 1987).

The Center for Education Statistics, in the reorganized Office of Educational Research and Improvement of the U.S. Department of Education, has initiated several studies to improve basic data collection activities. In response to the need for better information on the characteristics of the teaching force in elementary and secondary schools, a study is being carried out by the Rand Corporation to redesign existing surveys concerning teachers. The redesign effort has the goal of providing better estimates of teacher supply and demand as well as better information on qualifications of teachers, job characteristics, and conditions for teaching. A pilot study of the new design, which will include surveys at the teacher, school, and district levels, was conducted in 1986, with full implementation scheduled for 1987.

The redesign of data collection on teachers may contribute to a plan for revising the Center for Education Statistics' collection of data on elementary and secondary education. Among other proposed new data collection activities are new assessments of cognitive learning of students at several grade levels. A new long-term study of the educational performance and occupational attainment of a national sample of students, the National Education Longitudinal Study (NELS), will begin in 1988.

The Center for Education Statistics is also working with the Council of Chief State School Officers to ensure that the common core of data reported by school systems to the states and by the states to the center is accurate and timely. The goals are to describe state collection of data elements currently contained in the common core of data, to consider and describe what elements might be added, and to recommend means for making the common core of data more comprehensive, comparable, and timely.

Because of the general dissatisfaction with current achievement tests, the National Science Foundation supported a project at the Educational Testing Service to develop better measures for assessing student knowledge and performance in science. A manual has been published for science and mathematics coordinators and teachers on exercises designed for hands-on assessment of such skills as classifying, observing and making inferences, formulating hypotheses, interpreting data, designing an experiment, and conducting a complete experiment (National Assessment of Educational Progress, 1987).

New Survey Data Applicable to Indicators

Several surveys being completed during the same period as this study will provide new data relevant to indicators of precollege science and mathematics education. A national survey of science and mathematics education is currently being conducted by the Research Triangle Institute with funding from the National Science Foundation. This survey of teachers and principals will produce nationally representative data on the condition of science and mathematics education in elementary and secondary schools. Included in the survey are questions on course offerings and enrollments, availability of facilities and equipment, instructional techniques, textbook usage, teacher background, and needs for in-service education. Since the survey will provide follow-up data to a similar 1977 survey, analysis of trends in science and mathematics education during the last decade will be possible.

New sources of data on teachers in elementary and secondary schools will be available from three studies supported by the Department of Education. First, the 1985 Public School Survey conducted by the National Center for Education Statistics (now the Center for Education Statistics) focused on the status and characteristics of public school teachers. A nationally representative sample of approximately 10,750 teachers and 2,800 school administrators was surveyed through mail questionnaires. The teachers were asked for information on their teaching activities, background and experience, conditions for teaching, specific teaching practices (e.g., homework assigned), salary level, and work outside teaching. The administrator questionnaire asked for information on school characteristics, staffing, teacher incentive plans, and conditions for teaching. Special analyses of science and mathematics teachers will be possible with the data.

The National Assessment of Educational Progress (NAEP) for the 1985–1986 school year included an assessment of science knowledge for students in grades three, seven, and eleven. A new feature of the 1985–1986 NAEP is a survey of the teachers of students in the assessment sample. The objective is to gather information on teachers' training and experience, classroom conditions, and teaching practices. The results will provide a new source of information on science in elementary and secondary schools.

The Department of Education is also supporting the fifth follow-up survey for the National Longitudinal Study of the senior class of

1972. This is the first follow-up that will include a special survey supplement for persons in the sample who are teachers or former teachers. This teacher supplement, partially supported by the National Science Foundation, will provide information on career patterns and decisions related to the teaching profession.

Another project that will provide data for indicators is the School Mathematics Monitoring Center, established at the University of Wisconsin with the support of the National Science Foundation. This center is collecting, analyzing, and reporting data on key indicators of change in mathematics instruction and performance. A major purpose of the center is to analyze the response of schools to current reform efforts and their progress in improving mathematics education over the coming years. Center products will include a data retrieval system that will be available to the National Science Foundation (NSF) and to other federal and state agencies and a report to NSF on the status of mathematics education in the United States.

Efforts to Develop Indicators of the Quality of Curriculum

One response of professional science associations to the need for better indicators of science and mathematics education has been to work on new standards for curriculum and instruction. A major long-term project, Project 2061, has been initiated by the American Association for the Advancement of Science (AAAS) to define essential learning in science and mathematics that should be attained by all high school graduates. AAAS is working with panels of scientists, mathematicians, and educators to establish the core elements of science and mathematics that should be learned in school and where in the curriculum and at what age levels these elements should be taught.

Several professional associations have developed goals for improving curriculum and instruction in specific subject areas. For example, a special committee of the American Chemical Society (1984) developed a set of recommendations and guidelines for quality chemistry education programs at the high school and college levels. The National Council of Teachers of Mathematics (1980, 1981a, 1981b) has recommended actions for improving the quality of education in mathematics, and the National Science Teachers Association (1983) has established standards for preparation and certification of teachers in science for kindergarten through grade 12. These recommendations from professional associations can provide a basis for schools, districts, or states to measure the quality of their programs in science or mathematics.

Another approach to improving assessment of the quality of curriculum and instruction in science and mathematics has been to develop "frameworks" for curriculum content. Some states, such as South Carolina and California, have developed frameworks for science for specific grades and courses. These frameworks have been developed in conjunction with efforts to develop statewide competency examinations to assess student learning. A curriculum framework defines the core concepts to be learned by each student and, thus, can be used as a standard for assessing the curriculum and program of a school in a given subject area.

REFERENCES

American Chemical Society
- 1984   Tomorrow: The Report of the Task Force for the Study of Chemistry Education in the United States. Washington, D.C.: American Chemical Society.

Burstein, Leigh, Baker, Eva L., and Keesling, Ward J.
- No date   Using State Test Data for National Indicators of Educational Quality: A Feasibility Study. Final Report. Available from the Center for the Study of Evaluation, University of California, Los Angeles.

National Assessment of Educational Progress
- 1987   *Learning by Doing.* Princeton, N.J.: Educational Testing Service.

National Commission on Excellence in Education
- 1983   *A Nation At Risk: The Imperative for Educational Reform.* Supt. of Doc. No. 065-000-00204-3. Available from the U.S. Government Printing Office. Washington, D.C.: U.S. Department of Education.

National Council of Teachers of Mathematics
- 1980   *An Agenda for Action.* Reston, Va.: National Council of Teachers of Mathematics.
- 1981a   *Guidelines for the Preparation of Teachers of Mathematics.* Prepared by the Commission on the Education of Teachers of Mathematics. Reston, Va.: National Council of Teachers of Mathematics.
- 1981b   *Priorities in School Mathematics.* Reston, Va.: National Council of Teachers of Mathematics.

National Science Board
- 1985   *Science Indicators: The 1985 Report.* NSB 85-1. Washington, D.C.: National Science Foundation.

National Science Board Commission on Precollege Education in Science, Mathematics, and Technology
- 1983   *Educating Americans for the 21st Century.* Washington, D.C.: National Science Foundation.

National Science Teachers Association
  1983 *Recommended Standards for the Preparation and Certification of Teachers of Science at the Elementary and Middle/Junior High School Levels.* Washington, D.C.: National Science Teachers Association.
Oakes, Jeannie
  1986 *Educational Indicators: A Guide for Policymakers.* OPE-01. Santa Monica, Calif.: Rand Corporation.
Panel on Statistics on Supply and Demand for Precollege Science and Mathematics Teachers
  1987 *Toward Understanding Teachers Supply and Demand: Priorities for Research and Development.* Interim Report. Available from the Commission on Behavioral and Social Sciences and Education. Washington, D.C.: National Academy Press.
Raizen, Senta A., and Jones, Lyle V., eds.
  1985 *Indicators of Precollege Education in Science and Mathematics. A Preliminary Review.* Committee on Indicators of Precollege Science and Mathematics Education, National Research Council. Washington, D.C.: National Academy Press.
Shavelson, Richard, McDonnell, Lorraine, Oakes, Jeannie, and Carey, Neil
  1987 *Indicator Systems for Monitoring Mathematics and Science Education.* R-357D-NSF. Santa Monica, Calif.: Rand Corporation.
U.S. Department of Education
  1985 *Indicators of Education Status and Trends.* Washington, D.C.: U.S. Department of Education.
  1987 *Elementary and Secondary Education Indicators in Brief.* IS 87-106. Washington, D.C.: U.S. Department of Education.

Appendix E
Coordination of Strategies for Collecting Data

Data collection strategies must be planned in such a way that there is effective coordination in conducting the activities to develop indicator variables, and in cooperating with other organizations, such as the Center for Education Statistics and the National Assessment of Educational Progress (NAEP), that are collecting similar data. The committee therefore recommends that the collection of data for indicator objectives and for other ongoing and planned data collection activities be coordinated to the maximum extent feasible and effective. Indefensible multiplication of surveys or excessive burden on individual respondents would increase the costs of the program and decrease response rates. Effective coordination can lead to reduced costs through possible reduction in sample sizes and to improvements in data quality through sharing of frame development and maintenance, sampling operations, and improved training and quality control in survey administration. The purpose of this appendix is to discuss some of the considerations in achieving such coordination. The specific approaches and designs should be developed by those assigned responsibility for the new activities.

Elements of data collection activities that affect the feasibility of coordination are:

- Target population
- Data elements

APPENDIX E

- Methods of data collection
- Frequency of data collection
- Sample design (related to data collection methodology and frequency)
- Time of year (related to data elements and time cycles within the school year)
- Respondent burden and its impact on the cooperation of sample respondents

Much of the thrust of the committee's recommendations is in the direction of developing new assessment tools and descriptive data. It is far from certain that they will require separate vehicles. However, coordination with other assessment and data collection activities may depend upon the feasibility of compromise between the parallel surveys and gradual shifting to the new tools. Along with the development and testing of data items, some attention must be given to logistical concerns.

The committee considers the assessment of student learning to be of primary importance. For this purpose, the committee suggests testing of students in three grade levels, for example, grades 4, 8, and 12. The committee also recommends that data for additional key indicators about the students and about their teachers be obtained, as well as data to provide several supplementary indicators. These key indicator variables, for example, might include:

- For students: Semesters of science and mathematics taken by students in the 12th grade; time per week spent on science and mathematics study by students in 4th and 8th grades.
- For teachers: Knowledge in subject matter they are expected to teach.

Collection of these additional data, linked to student learning, can serve two purposes: (1) to provide descriptive statistics about students, teachers, and schools with regard to the distribution of factors linked to student learning and (2) to help understand and explain differences in student achievement.

For the first purpose, linkages between individual students and their teachers are not necessary. Samples of schools, teachers, and students (and possibly their parents) not necessarily linked to the sample of students selected for testing at specified grades can be used to provide general descriptive statistics. Consideration should be given to coordinating these samples with the Elementary/Secondary Integrated Data Systems (ESIDS) program being developed by the

Center for Education Statistics. Preliminary specification of sample sizes for schools, students, and teachers could be based on design parameters used by the Center for Education Statistics. In any event, however, steps should be taken to ensure that the data would be comparable to corresponding data from schools and teachers associated with the sample of students selected for testing.

For the second purpose, the committee recommends that the sample be tied to students classified by race or ethnicity, gender, grade level or age, socioeconomic status, type of community, and region or state. It must be recognized that, realistically, it will not be possible with the levels of effort now represented by ESIDS or NAEP, for example, to provide enough data for all cross-classifications of these variables. It will not even be possible to provide for, say, cross-classification of race or ethnicity and gender with equal precision in every cell. One design approach would be to set limited goals but establish designs that could readily be extended; for example, a design for national data that could be expanded to provide state-level data.

Testing of students might be coordinated with NAEP, depending on the time needed. NAEP now requires about an hour of student time, and this could not be reduced substantially without jeopardizing it. Coordination would then depend on the feasibility of adding to the time per student or of sampling additional students.

Certain new activities are likely to be special efforts, although in each case it would be desirable to associate them with an organization having related data collection or analysis responsibilities. Among them are the following:

Recommendation by Committee	Existing Organization/Activity
Salary survey	U.S. Census Bureau
Federal support of science and mathematics education	National Science Foundation
Support of scientific bodies	
Observation of classroom processes	International Association for the Evaluation of Education Achievement National Assessment of Educational Progress Elementary and Secondary Integrated Data Systems Linkage to developing teacher evaluation programs, in the R&D phase
Constructions of curriculum frameworks	American Association for the Advancement of Science Mathematical Sciences Education Board National Council of Teachers of Mathematics

Index

A

Accuracy of information
  in achievement tests on science, 178
  in curriculum, 13, 125, 137-138, 139, 140
Achieved curriculum, 121
Achievement tests
  multiple-choice, as indicator of student learning, 40-50
  on science, 48-49, 175-180
    high-quality, characteristics of, 178-179
    panel on review of, 176, 180
    scores on, and socioeconomic background of students, 31-32
Activities
  in-school
    of students, 77-79, 81-82
    of teachers, 105-108
  out-of-school
    of students, 79-80, 82-83
    of teachers, 103-105
Actual or implemented curriculum. *See* Implemented curriculum
Agencies
  federal, financial support of science and mathematics education, 4, 14, 144-149
  local education
    financial support of science and mathematics education, 143-144, 148-149, 186-187
    summary of meetings with, 189-196
  national scientific, support of science and mathematics education, 4, 14, 149-151
  state education. *See* State education agencies
Aggregation of data, 31-33, 199
  and ecological fallacy, 31-32
  inconsistent, 33
  levels of, 31, 32
  self-selection in, 33
American Association for the Advancement of Science, 120, 149, 203
American Chemical Society, 203
  support of science and mathematics education, 149-150
American College Testing Program, 33
American Geological Institute, 150
American Institute of Biological Sciences, 150
American Institute of Physics, 149

American Mathematical Society, 149
Anthropology, in study of education, 23
Army Alpha tests, in World War I, 42
Assessment Center of the Council of Chief State School Officers, 39, 66
Attitudes of students toward science and mathematics, 75–76, 84–85
  meetings with representatives of education agencies on, 187, 194–195
  participants in colloquium on, 173
  research and development recommendations on, 85
Attribution of success, and sense of fate control, 86–87
Automaticity of processing speed, measures of, 56
Autonomy sense of students, 88

B

Balanced incomplete block design of tests, 50
Behavior
  and outcome of schooling, 22, 23–26
  of students, 2, 3, 7–9, 29, 73–89. *See* Teachers,
  time use studies on; Time use studies
Biases, in panel assessments, 38
British Assessment of Performance Unit Series, 5, 53, 64

C

California
  assessment of student attitudes in, 187
  state guidelines on curriculum in, 127, 128, 140, 182–183, 204
  student testing in, 185
California Assessment Program, science content of, 177
Canada, curriculum guidelines in, 133
Carnegie Forum on Education and the Economy, 91, 93–94, 138

Center for Education Statistics, 39, 69, 201, 202, 206, 208
Center for Policy Research in Education, 199
Center for the Study of Evaluation, 199
Certification of teachers, testing for, 109, 110
Civic literacy in mathematics, 21
Classroom instruction time on science and mathematics. *See* Instructional time on science and mathematics
Coaching of students on achievement tests, 45–46
  computerized systems in, 60
Collection of data on indicators, 35–39. *See also* Data on indicators, collection of
College Board Advanced Placement Tests, 5, 64
College education of teachers
  collection of data on, 95–96, 102
  entrance examinations in, 109
  as indicator of teaching quality, 10, 95–96, 100, 102, 109
  and salaries, compared with other college graduates, 113–117, 118
College Placement Council, data on salaries of teachers, 117
Colloquium on indicators of precollege science and mathematics education, 171–174
Comparable data, collection of, 35–36
  on salaries of teachers and other occupations, 113–117
Compensatory education for disadvantaged children, 146–147
Competence, student perception of, 88
Competency testing
  of students, 46–47
  of teachers, 97–98, 101, 102, 108–110
Comprehensive Tests of Skills, science content of, 17
Computers
  affecting mathematics curriculum, 131–132

INDEX 211

in assessment of problem-solving
  skills, 60–61
in assessment of student learning,
  52, 53–54, 184, 185
on physical laws, 59–60
on procedural knowledge, 60–61
and processing speed, 56
improving student learning, 53–54,
  60
in simulations of scientific
  experiments, 54, 59–60
Conceptual knowledge, assessment
  of, 55–61, 63
of adult population, 71
on internal representations of
  problems, 58–60
on organization of knowledge in
  memory, 57–58
Conference Board of the
  Mathematical Sciences, 128, 131
Connectedness, student perception
  of, 88
Connecticut, study of newly hired
  teachers in, 100
Constraints influencing teacher and
  student behavior, 24–26
in curriculum, 119, 125
Content coverage of curriculum, 12,
  13, 124, 125, 127–135
depth of, 125, 136–137, 139, 140
expert panels in assessment of,
  132–133
frequency of assessment, 134, 135
on mathematics, 130–132
state guidelines on, 133
teacher reporting of, 13, 134, 135
in tests, 133
on science, review of, 175–180
in textbooks, 128–129, 133
Coordination of data collection
  strategies, 38–39, 206–208
Correlation coefficients, aggregation
  of data affecting, 31–32
Council of Chief State School
  Officers, 39, 66, 140–141, 200,
  201
Course enrollment data, 2, 7, 77–78,
  79
recommendations on, 81
Creative thinking, exclusion of, in
  multiple-choice tests, 43–44

Cultural literacy in mathematics, 22
Current projects on indicators,
  197–205
Curriculum, 2, 3, 12–14, 119–142
accuracy of information in, 13, 125,
  137–138, 139, 140
achieved, 121
computers affecting, 131–132
content coverage of, 12, 13, 124,
  125, 127–135
definition of, 120–123
depth of topic treatments in, 125
indicators of, 136–137, 139, 140
development of indicators on,
  203–204
expert panel in assessment of,
  132–133, 137, 139, 140
frameworks for, 12, 127–132, 182,
  191, 192, 204
national, 141, 182
recommendations on, 134–135
frequency of assessment of, 134, 135
by grade clusters, 122–123, 128,
  129, 135
homework time as measure of, 127
implemented or actual. See
  Implemented curriculum
incentives and constraints in, 119,
  125
instructional time as measure of,
  126–127
intended. See Intended curriculum
key indicators on, 12–13, 135, 140
on mathematics
  frameworks for, 130–132
  state guidelines on, 128, 129
meetings with representatives of
  education agencies on, 182–183,
  191–192
models of excellence in, 139
number of science and mathematics
  courses taken as measure of, 126
pedagogical quality of, 13, 14, 125,
  138, 139, 140
research and development
  recommendations on, 12, 13, 14,
  134–135, 139, 140
response of teachers to changes in,
  106, 108
role of teachers in planning and
  shaping, 26

on science
frameworks for, 132
state guidelines on, 132
spiraled, 125
state guidelines on, 12, 13, 121, 127, 128–129, 132, 135, 182–183, 204
depth of topic treatments in, 137
implications of indicators for, 140–142
teacher reporting of, 13, 134, 135
tests in, 122, 128–129, 135, 141–142
content coverage of, 133, 175–180
depth of topic treatments in, 137
scientific accuracy of, 137
tests influencing, 46–48
textbooks as part of, 12, 13, 121–122, 135, 141
depth of topic treatments in, 137
measures of content coverage in, 128–129, 133
scientific accuracy of, 137
types of indicators on, 124–126
users of indicators on, 123–124

D

Data on indicators
aggregation of, 31–33, 199
collection of, 35–39
on college education of teachers, 95–96, 102
of comparable and unexpected information, 35–36
coordination of strategies in, 38–39, 206–208
depth of information in, 35–36
expert panels in, 36–38
on federal financial support of science and mathematics education, 145–149
frequency of, 36
on implemented curriculum, 134
improvements in, 200–201, 202
multiple-choice achievement tests in, 40–50
on salary of teachers, 115–116
on scientific and mathematical literacy of adults, 69–71
on subject-matter knowledge of teachers, 99–100
on time use of teachers, outside of classroom, 104
on working conditions in schools, 111–112
Department of Education
current projects funded by, 199, 202
report on indicators, 198
Disadvantaged children, compensatory education for, 146–147
Discretionary tasks in learning science and mathematics, 88
District level of data aggregation, 31, 32

E

Ecological fallacy, in aggregation of data, 31–32
Economics, in study of education, 23
Education for All Handicapped Children Act, 146
Education Consolidation and Improvement Act, 146
Education for Economic Security Act of 1983, 146
Educational Testing Service, 77, 198–199, 201
Elementary and Secondary Education Act of 1965, 46
Elementary/Secondary Integrated Data Systems program, 207, 208
Engagement in learning science and mathematics, 88
Enrollment data on science and mathematics courses, 2, 7, 77–78, 79, 81
Equality of educational opportunity, 23
in curriculum, 141
Essay tests, as learning indicator, 51–53
Examinations. *See* Testing
Excellence in science and mathematics
in curriculum, models of, 139
distribution of, 26
Expectations of students, 88
Expert panels, in assessment procedures, 36–38

INDEX 213

biases of, 38
on curriculum, 132–133, 137, 139, 140
rater variability in, 37–38
on science achievement tests, 176, 180
validity and reliability of, 38

F

Fate control, sense of, 86–87
Feasibility of indicators, 35
Federal financial support of science and mathematics education, 4, 14, 144–149
  agency budgets in, 146–147, 148
  categories of funding in, 146, 147, 148
  collection of data on, 145
  supplementary indicator on, 148
Financial support of mathematics and science education, 143–151
  federal, 4, 14, 144–149
    budgets of agencies, 146–148
  local, 143–144, 148–149, 186–187
  meeting with state education agencies on, 186–187
  participants in colloquium on, 174
  from scientific organizations, 149–151
    budgets, 149–151
  state, 148–149, 186–187
  traditional measures of, 143–144
Florida, student testing in, 46, 185, 193
Frameworks, curriculum, 12, 127–132, 182, 191, 192, 204
  development of, 129–132
  on mathematics, 130–132
  national, 141, 182
  recommendations on, 134–135
  on science, 132
France, curriculum in, 133, 182
Free-response tests
  in assessment of adult scientific and mathematical literacy, 71, 72
  as indicator of student learning, 51–53
    compared with multiple-choice tests, 43–44
    recommendations on, 64

validity of, 62–63
Frequency of data collection, 36

G

Global assessment procedures of student learning, 51–54
Government
  federal financial support of science and mathematics education, 4, 14, 144–149
  state education agencies. *See* State education agencies
Graduate Record Examinations, 52, 199
Graduation requirements, state guidelines on, 187
Great Britain, curriculum quality in, 182

H

Habits of mind, scientific and mathematical, 18–19, 75, 76, 85–89, 187
  research and development recommendations on, 89
Handicapped children, education of, 145, 146
Hands-on instruction, 5, 53, 64
  quality of teaching in, 107, 108
High School and Beyond Longitudinal Survey, 177, 199
Holmes Group Consortium, 91, 93, 138
Homework on science and mathematics
  teacher correction of and feedback on, 106
  time spent on, 4, 8, 80, 82
  as curriculum indicator, 127
  recommendations on, 82
Human affairs, role of science in, 19–20
Hypothesis formulation testing, 5, 51–52, 64

I

Ideational fluency, 44

Illinois
  mathematics curriculum in, 128
  study of newly hired teachers in, 100
Implemented curriculum, 121, 122
  assessment of, 122, 129, 134, 135
  collection of data on, 134
  compared with mandated curriculum, 25–26
  scientific accuracy of, 137–138
Incentives influencing teacher and student behavior, 24–26
  in curriculum, 119
  salary as, 113–117
Indicators of science and mathematics education, 27–39
  aggregation of data on, 31–33
  collection of data on, 35–39
  current projects on, 197–205
  definition of, 27–29
  feasibility of, 35
  interpretation of, 29–34
  key, 2–3, 29
  scale of, 33–34
  supplementary, 2, 4, 29
  users of, 34–35, 123–124
  variables affecting, 30
In-school activities
  of students, 7–8, 77–79, 82
  recommendations on, 81–82
  of teachers, 105–108
Instructional time on science and mathematics, 7–8, 78–79, 81–82
  as behavior indicator, 7–8
  as curriculum quality indicator, 126–127
  minutes of, 78, 81–82
  student use of, 78–79, 82
  teacher use of, 105–106, 110–111
  as teaching quality indicator, 92
Intended curriculum, 121–122
  assessment of, 121–122, 127–129, 135
  compared with actual curriculum, 25–26
  depth of topic treatments in, 137, 140
  scientific accuracy of, 137, 140
Internal representations of problems, assessment of, 58–60

International Association for the Evaluation of Educational Achievement, 5, 53, 64, 85, 124, 134
  science content in tests of, 177
Interpretation of indicators, 29–34
Item banks for student tests, 185–186, 192–193
Item-response theory on multiple-choice tests, 48, 50

J

Japan
  curriculum quality in, 133, 182
  mathematics education in, 124

K

Key indicators of science and mathematics education, 2–3, 29
  curriculum quality, 12–13, 135, 140
  student behavior and learning, 2–3, 6–9, 65, 72, 81–82
  teaching quality, 9–10, 11–12, 102, 118
Knowledge of subject-matter of teachers. *See* Teachers, subject-matter knowledge of

L

Laboratory facilities of school, 111, 112, 118
  influencing student and teacher behavior, 25
Leadership of scientific organizations in education, 4, 14, 149–151
Learning in science and mathematics, indicators of, 2, 3, 4–7, 40–72
  theoretical basis of, 23
Literacy, scientific and mathematical, 2, 3, 6–7, 15–22, 67–72
  collection of data on, 69–71
  conceptual knowledge in, 71
  dimensions of, 16–22
  importance of, 67
  levels of, 20–21

mathematical, 20–22
meetings with representatives of education agencies on, 187, 194
participants in colloquium on, 173–174
recommendations on assessment of, 72
scientific, 16–20
target populations for assessment of, 69
Local education agencies
  financial support of science and mathematics education, 143–144, 148–149, 186–187
  results of meeting with representatives of, 189–196
Long-term memory, organization of knowledge in, 57–58

M

Mandated curriculum, compared with actual curriculum, 25–26
Materials and supplies, instructional, 4, 111–112, 118
  as curriculum quality indicator, 12, 13
  influencing teacher and student behavior, 25, 26
  meetings with representatives of education agencies on, 186–187, 193
  as teaching quality indicator, 11, 25, 111–112, 118
Mathematical Association of America, 149
Mathematical Sciences Education Board, 128
Matrix sampling, 49–50, 185
Meetings with state and local education agencies, summaries of, 181–196
Memory
  organization of knowledge in, assessment of, 57–58
  retrieval of information from, assessment of, 58
Minnesota, student testing in, 185
Minority science and mathematics teachers, 109

Models of schooling
  education production function, 23
  input-output, 74–75, 91–93
  process-product, 91–92
  students and teachers as key actors, 23, 91–92
Motivation of students, 75–76, 84–85
  meetings with representatives of education agencies on, 187, 194–195
  participants in colloquium on, 173
Multiple-choice tests, 40–50
  balanced incomplete block design of, 50
  coaching of students on, 45–46
  comparison of results on over time, 50
  criticisms of, 41–49
  as economical measure, 49
  free-response tests compared with, 43–44
  influence on curriculum, 46–48
  item sampling of, 49–50
  lack of creative thinking in, 43–44
  as learning indicator, 40–50, 63
  real-life problems compared to, 44
  science content of, 48–49, 175–180
  statistical analysis of, 50
  theoretical basis of, 48
Museum visits, out-of-school, 80, 82

N

National Academy of Sciences, 46, 149
National Assessment of Educational Progress, 30, 199, 202, 206, 208
  assessment of student attitudes, 84, 85
  on decline of student performance, 46, 47
  surveys of, 69, 77
  test materials of, 5, 50, 53, 64, 177
National Center for Education Statistics, 202
National Commission on Excellence in Education, 197
National Council of Teachers of Mathematics, 128, 203
National curriculum frameworks, 141, 182
National Defense Education Act, 146

National Education Longitudinal Study, 101, 201, 202-203
National leadership in science and mathematics education, 4, 14, 149-151
National Longitudinal Study of the High School Class of 1972, 101
National Research Council, 200
National Science Board Commission on Precollege Science, Mathematics, and Technology Education, 197, 198
National Science Foundation, 14, 146, 148, 199, 200
  Rand Corporation report for, 198
  research projects funded by, 202, 203
  on student testing, 201
National Science and Mathematics Assessment Resource Center, 6, 65-66
National Science Resources Center, 46
National Science Teachers Association, 46, 203
New York
  curriculum guidelines in, 121, 183
  study of newly hired teachers in, 100
New York State Regents, science content of, 176, 177
North Carolina
  student testing in, 185
  teachers leaving teaching in, 100
Northwestern Endicott Report, 117

O

Oregon
  science curriculum in, 132
  student testing in, 185
Organization of knowledge in memory, assessment of, 57-58
Organizations, scientific, support of science and mathematics education, 4, 14, 149-151
Out-of-school activities
  of students, 8-9, 79-80
  recommendations on, 82-83
  of teachers, 103-105, 107

P

Panels of experts, in assessment procedures. *See* Expert panels, in assessment procedures
Pattern recognition, assessment of, 56-57
Pedagogic quality of curriculum, 13, 14, 125, 138, 139, 140
Practical literacy in mathematics, 21
Precollege education, use of term, 195
Problem solving
  computerized assessment of, 60-61
  free-response tests of, validity of, 62-63
  internal representations of problems in, 58-60
  as learning indicator, 5
  procedural knowledge in, 60-61
  in real-life, compared with multiple-choice tests, 44
  "think aloud" method in assessment of, 59
Procedural knowledge, assessment of, 60-61
  computers in, 60-61
Process-product studies of teaching quality, 91, 92-93
Processing skills, assessment of, 51, 55-61
  on internal representations of problems, 58-60
  on organization of knowledge in memory, 57-58
  on pattern recognition, 56-57
  on procedural knowledge, 60-61
  on retrieval of information from memory, 58
  on speed of processing, 55-56
Professional teachers, 94
Project TALENT, 31
Project 2061, 203
Psychology, educational, 23, 74
Public policy decisions
  mathematical concepts in, 21
  scientific concepts in, 19-20

Q

Questioning techniques of teachers, in classroom, 107, 108
Questionnaires in data collection, closed-ended, 35–36

R

Rand Corporation, 198, 201
Rater variability in panel assessments, 37–38
Reliability
  of essay tests, 51
  of panel assessments, 38
Representations of problems, internal, assessment of, 58–60
Representatives of education agencies, summary of meetings with, 181–196
Research Triangle Institute, 202
Reserve pool of experienced teachers, 100
Resources for teaching science and mathematics, 4, 11–112, 118
  as curriculum quality indicator, 12, 13
  influencing teacher and student behavior, 25, 26
  meetings with representatives of education agencies on, 186–187, 193–194
  as teaching quality indicator, 11, 25, 111–112, 118
Response latencies, in measurement of processing speed, 55–56

S

Salary of teachers, 3, 11–12, 113–117, 118, 187
  and career decisions, 113
  collection of data on, 115–116
  compared with other occupations, 11–12, 24, 113–117, 118
  and second jobs, 103
  and time use outside classroom, 103
Scale of indicators, interpretation of, 33–34
Scholastic Aptitude Tests, 199
  coaching of students on, 45
  interpretation of scores on, 33
School level of data aggregation, 31, 32
School Mathematics Monitoring Center, 203
School Mathematics Study Group, 149
Schooling outcomes, 22–26
  behavior of students and teachers affecting, 22, 23–24
  and distribution of excellence, 23, 26
  incentives and constraints affecting, 22, 24–26
Scientific method, 19
Scientific organizations, support of science and mathematics education, 4, 14, 149–151
Scientific world view, 16–17
Simulations of scientific experiments, computer-aided, 54, 59–60
Society, importance of science and mathematics in, 19–20, 21, 80, 83
Socioeconomic background, and achievement test scores, 31–32
Sociology, in study of education, 23
South Carolina
  curriculum guidelines in, 127, 140, 182, 204
  teacher testing in, 183
Speed of processing, assessment of, 55–56
  pattern recognition in, 56–57
Spiraled curriculum, 125
State education agencies
  curriculum guidelines of, 12, 13, 121, 127, 135, 182–183, 204
    on content coverage, 133
    depth of, 137
    implications of indicators for, 140–142
    on mathematics, 128, 129
    on science, 132
  financial support of science and mathematics education, 148–149, 186–187
  graduation requirements of, 187
  student learning indicators for, 66–67

summary of meetings with, 181–188
teaching quality indicators for,
    108–111
State level of data aggregation, 31,
    32
Student learning
  assessment of, 51–66
  computerized assessment of, 52–54,
    56, 59–61
  computers improving in classroom,
    53–54, 60
  engagement in, 88
  global assessment of, 51–54
  indicators of, 2–7, 23, 40–72
  key indicators of, 2–3, 6–8, 65,
    81–82
  research and development
    recommendations on, 8–9
  and behavior, 64–66, 82–83
  supplementary indicators on, 4, 8,
    82
Students, 40–89
  aggregation of data on, 31–33
  attitudes toward science and
    mathematics of, 75–76, 84–85
  meetings with representatives of
    education agencies on, 187,
    194–195
  participants in colloquium on, 173
  research and development
    recommendations on, 85
  autonomy sense of, 88
  behavior indicators on, 2, 3, 7–9,
    29, 73–89
  center for production and
    distribution of assessment
    materials on, 6, 65–66
  competence of, 88
  conceptual knowledge and
    processing skills of, 51, 55–61
  course enrollment data on, 2, 7,
    77–78, 79
    recommendations on, 81
  as determinant of schooling
    outcome, 22, 23–24
  incentives and constraints
    affecting, 24–26
  expectations of, 88
  fate control, sense of, 86–87
  frequency of assessment, 65

habits of mind of, scientific and
    mathematical, 18–19, 75, 76,
    85–89, 187
  hands-on procedures in instruction
    of, 5, 53, 64, 107, 108
  homework on science and
    mathematics of, 4, 8, 80, 82,
    106, 127
  input-output model on, 74–75
  in-school activities of, 7–8, 77–79,
    82
    recommendations on, 81–82
  as key actors, 73–76
  new methods of assessment, 51–66
  out-of-school activities of, 8–9,
    79–80
    recommendations on, 82–83
  perception of connectedness, 88
  problem-solving skills of, 5, 44,
    58–61, 62
  testing of. *See* Testing, of students
Subject-matter knowledge
  of students, indicators on, 2, 3,
    4–7, 40–72
  of teachers. *See* Teachers,
    subject-matter knowledge of
Supplementary indicators of science
  and mathematics education, 2,
  4, 29
  federal financial support, 148
  student behavior and learning, 4, 8,
    82
  teaching quality, 4, 10–11, 102,
    107–108, 118
Supply and demand for teachers,
  100, 200
  and reserve pool of experienced
    teachers, 100

T

Teachers, 3, 9–12, 90–118
  collection of data on, improvements
    in, 200, 201, 202
  college education of, 10, 95–96,
    100, 102, 109, 113–117
  computers in instruction methods
    of, 53–54, 60
  curriculum changes affecting, 106,
    108

as determinant of schooling
outcome, 22, 23-24
incentives and constraints
affecting, 24-26
early home and school experiences
affecting, 101-102
encouraging intellectual curiosity,
184, 190
hands-on instruction methods of, 5,
53, 107, 108
implications of indicators on, for
state education agencies
108-111
input-output studies of, 91, 92, 93
instructional materials and supplies
used by. *See* Materials and
supplies, instructional
interpretation of indicators on, 29
as key actors, 90-94
key indicators on, 9-10, 11-12, 102,
118
leaving teaching after one year, 100
literature on, 92-93
meetings with representatives of
education agencies on, 183-184,
189-191
outcomes-based model of, 190
participants in colloquium on,
172-173
pedagogic knowledge of, 138
process-product studies of, 91,
92-93
professional, 94
quality of curriculum delivery by,
184
questioning techniques of, 107, 108
reporting on content coverage of
curriculum by, 13, 134, 135
research and development
recommendations on, 10, 11,
103, 108
resources in school affecting, 4,
111-112, 118
salary of. *See* Salary of teachers
subject-matter knowledge of, 95,
96-100, 101, 102, 108-110
and accuracy of curriculum,
137-138
assessment of, 183, 190
collection of data on, 99-100

implications for state education
agencies, 108-110
of newly hired teachers, 99, 100
recommendations on testing of,
102
research needed on, 101
supplementary indicators on, 4,
10-11, 102, 107-108, 118
supply and demand for, 100, 200
and reserve pool of experienced
teachers, 100
testing of. *See* Testing, of teachers
tests of student learning in
evaluation of, 41-42
time use studies on, 10-11, 103-108
in classroom, 92, 105-106,
110-111, 126-127
as curriculum quality indicator,
126-127
outside classroom, 103-105, 107
recommendations on, 107-108
research needed on, 106-107
working conditions in school
affecting, 111-118
years of experience of, 92
Tennessee
student testing in, 185
teacher evaluation in, 108, 184
Testing
of students, 5-6, 40-50
accuracy of information in, 137,
178
balanced incomplete block design
of tests in, 50
computerized, 184, 185
content coverage of, 133, 175-180
creative thinking in, 43-44
criticisms of, 41-49
in curriculum, 122, 128-129, 135,
141-142
depth of topic treatments in, 137
essay tests in, 51-53
financial support for, 148
free-response tests in, 43-44,
51-53, 62-63, 64
frequency of, 6
ideational fluency in, 44
instructional applications of,
41-42
item banks for, 185-186, 192-193
item response theory on, 48, 50

meetings with representatives of education agencies on, 184–186, 192–193
multiple-choice tests in. *See* Multiple-choice tests
purposes of, 41–42
science content in, 175–180
statewide, 185
virtues of, 49–50
of teachers, 9–10, 97–98, 101, 102, 108–110, 137–138, 183, 191
for certification, 109, 110
collection of data on, 99–100
frequency of, 10, 99, 102
implications for state education agencies, 108–110
recommendations on, 102
Tests of Achievement and Proficiency, science content of, 177
Texas, curriculum guidelines in, 140
Textbooks, in curriculum, 12, 13, 121–122, 135, 141
content coverage of, 128–129, 133
depth of topic treatments in, 137
scientific accuracy of, 137
"Think aloud" method in assessment of problem-solving skills, 59
Time use studies, 4, 7–8, 36
on homework time on science and mathematics, 4, 8, 80, 82,
on instructional time on science and mathematics. *See* Instructional time on science and mathematics
on teachers, 10–11, 103–108. *See also* Teachers, time use studies on
Timing of data collection, 36

U

Users of indicators, 34–35
on curriculum quality, 123–124

V

Validity
of panel assessments, 38
of student learning indicators, 62–63

Variability in panel assessments, 37–38
Variables affecting interpretation of indicators, 30
Videotaping of problem-solving behavior, 59
Virginia
curriculum guidelines in, 127
student testing in, 185

W

West Germany, curriculum guidelines in, 133, 182
Wisconsin, mathematics curriculum in, 128
Working conditions in schools, 4, 25, 26, 111–118
data collection on, 111–112
as indicator of curriculum quality, 12, 13
as indicator of teaching quality, 11, 25, 111–118
meetings with representatives of education agencies on, 186–187, 193–194
salaries in, 113–117, 118
World view, scientific, 16–17

Z

Zoo visits, out-of-school, 80, 82